The Minds of Birds

Number Twenty-three:
The Louise Lindsey Merrick
Natural Environment Series

THE
Minds
OF
Birds

by Alexander F. Skutch

Illustrations by Dana Gardner

TEXAS A&M UNIVERSITY PRESS
College Station

Library of Congress Cataloging-in-Publication Data

Skutch, Alexander Frank, 1904–
    The minds of birds / by Alexander F. Skutch ; illustrated by
Dana Gardner.
        p.   cm.— (The Louise Lindsey Merrick natural
environment series ; no. 23)
    Includes bibliographical references and index.
    ISBN 0-89096-671-0 (cloth); 0-89096-759-8 (pbk.)
    1. Birds—Psychology. 2. Birds—Behavior. 3. Animal
intelligence. I. Title. II. Series.
QL698.3.S553   1995
598.251—dc20

To Liliana Halder and Sergio Saborío
with gratitude and affection

Come tell us, O tell us,
    Thou strange mortality!
What's *thy* thought of us, Dear?—
    Here's *our* thought of thee.

Alack, you tall angels,
    I cant think so high!
I cant think what it feels like
    Not to be I.

Come tell me, O tell me,
    My poet of the blue,
What's *your* thought of me, Sweet?—
    Here's *my* thought of you.

                    Francis Thompson

Contents

Illustrations

Preface

WE DELIGHT in the beauty of birds; we enjoy their songs; we admire their nests, so diverse and skillfully made without hands; we applaud their devoted care of their eggs and young, equaled by few other creatures; we wonder how they find their way over vast expanses of trackless oceans, forests, and deserts; but we tend to disparage their mentality. We may make exceptions for crows and jays, which may merely be more aggressive than other birds, and for parrots because some of them can repeat human words; but, on the whole, we consider birds stupid. Chickens rushing across the highway just ahead of a speeding car, birds fighting their reflections in a mirror or a shiny hubcap, birds injuring or killing themselves by flying against window glass or screens, and other common observations reinforce our opinion of their low intelligence. To say that someone "has a mind like a bird" is the equivalent of calling that person a dunce.

When we reflect that people, all classified in the same species, vary immensely in intelligence, we should expect considerable diversity in mental qualities among the approximately nine thousand species of birds and even appreciable difference among individuals of the same species; but, viewed widely, their mental qualities, too often undervalued, command respect. From diverse sources, we have good evidence that birds are far from stupid. In solving the problems that researchers give them in laboratories, their performances compare favorably with those of nonhuman mammals. They can be taught to count up to six or eight and to perform various tasks. But more than in these artificial situations, in their natural habitats they reveal their capacities. Their sense of space, demonstrated for example in finding a cache of food hidden in the ground months earlier and in migrating thousands of miles to exactly the same winter or summer territory as they have previously occupied, excites our wonder. Their ability to recognize individuals of their own kind, often bafflingly similar to us, and even to single out a known human in a crowd, is evidence of keen perceptions. Their interest in sounds, and the ability of many birds to repeat them, has been regarded by students of bird song as a major manifestation of their intelligence. In the past few decades, ornithologists have investigated avian social systems, exemplified by cooperative breeding arrangements exceeded in complexity only by those of humans. Many of these accomplishments depend upon memory—of places, individuals, or sounds—which, contrary to a widespread view, is well developed in birds.

Observations that reveal the quality of birds' minds, whether made in laboratory or field, are widely scattered in scientific publications as well as in books and periodicals intended for the general reader. To reach a fair estimate of the many-faceted minds of birds, one must survey a number of these reports, as I have tried to do in nontechnical language in this book. What emerges from such a comprehensive view is profound respect for the mental capacities of birds, and wonder that heads so small could hold the brains that are the foundation of their abilities.

When we have learned all that overt behavior reveals about the minds of birds, searching questions about their inner or psychic lives remain unanswered. How keenly aware are they of what they see, hear, and do? Do they enjoy singing the songs that delight us? Do they feel affection for the mates with whom they closely cooperate in rearing their broods, and whom, especially when permanently resident, they accompany throughout the year? Are they emotionally attached to the young they faithfully nurture? Do we correctly interpret the cries of distress that many utter when their nests are or appear to be jeopardized? Does an aesthetic sense influence a female's selection of a male elegantly adorned in nuptial attire, as in birds of paradise, hummingbirds, manakins, and many others? In short, in what measure do birds, whose habits suggest the possibility of a rich psychic life, enjoy such a life?

These questions have scientific, moral, and philosophic importance. To know their answers would immensely enrich our natural history and enlighten our studies of avian behavior. For an ethic that looks beyond humankind, they are of first importance. If, as some have held, nonhuman creatures are organisms with little feeling, we might treat birds as we do beautiful works of art or delicate artifacts such as watches and television sets. If we believe that birds enjoy and suffer somewhat as we do, we have stronger motives for protecting them. Our attitude toward evolution, and indeed toward the universe, can be strongly influenced by our estimate of the psychic lives of the creatures that share Earth with us. To believe that as evolution increases the complexity of organisms, it enhances their capacity for feeling, knowing, and enjoying mitigates and cheers our view of an undeniably harsh process, in which many of its learned interpreters find no purpose. We feel more comfortable in a world enlivened by beings akin to us in mind, as in anatomy and physiology, than in one in which we are lonely exceptions. Our whole attitude toward life, our religion in the profoundest sense of the word, is subtly colored by our worldview.

In this attempt to reach a fair estimate of avian mentality, I take account of field observations as well as the researches of psychologists in their laboratories. The former are often disparaged by professional biologists as "anecdotal" because they cannot be repeated or treated statistically. Birds are creatures of habit whose innate behaviors are so adequate for all the contingencies of their lives in natural surroundings that they rarely need to deviate from them. They do not often act in ways that clearly reveal innovative intelligence. But occasionally a once-in-a-lifetime observation of a bird doing something evidently not programmed in its genes gives us a precious glimpse of avian mentality and, if made by a competent and honest observer, is of inestimable value to students of bird behavior.

Soon after I began to study birds, well over sixty years ago, I began to wonder about their psychic lives. Often, in the excitement of learning about little-known birds, I forgot this deeper question; but ever and again it would recur to me, for I regarded their feelings and thoughts as the most important, but unfortunately the most baffling, aspect of their lives. After watching and reading about birds for so many years, I am far less certain of what goes on inside their heads than of what they visibly do, but I have reached some tentative conclusions that seem worth publishing, as a step toward a deeper understanding of the life around us. Thus, this book is the culmination of a lifelong quest. In it I treat of some large subjects—such as cooperative breeding, sexual selection and the aesthetic sense, migration, the structure of the avian brain and sense organs—with no pretension to thoroughness, but only so far as they might throw light on birds' minds. For readers who might wish to pursue any of these subjects in depth, I give references that should be helpful. In the text, I use the birds' common names; their scientific equivalents are given in the index.

In these days when birds nearly everywhere are decreasing alarmingly in abundance under the pressures of an already excessive and too rapidly increasing human population, too greedy for limited resources, they need all the protection that we can give them. To have good reasons for believing that they are not merely beautiful, songful organisms that increase our enjoyment of nature while they protect our forests and plantations from devouring insect hordes, but are feeling creatures whose own capacity for enjoyment increases the total value of life on Earth—this should intensify our determination to protect them, not only as species but also as individuals that cling to their lives as we do to ours.

I fear that critical readers will accuse me of anthropomorphism. In my student days, anthropomorphism was one of the most flagrant of scientific heresies, only a little less heinous than the unforgivable sin of falsifying one's observations or data. Anthropomorphism is the attribution to nonhuman creatures of behavior or mental qualities that we are pleased to consider uniquely human. Nevertheless, when we recognize that we can hardly imagine any psychic state that we do not from time to time experience in ourselves, the rigid avoidance of anthropomorphism might exclude the possibility of attributing to nonhuman animals any psychic life at all. Unfortunately, the uncritical ascription of human sentiments or activities to other animals, as by certain popular writers on natural history, has made serious naturalists recoil too strongly in the opposite direction. Birds are more likely to experience basic emotions, including fear, anger, love, or affection for offspring, than to be capable of such more complex sentiments as pride, shame, or remorse, sometimes carelessly attributed to them.

Anthropomorphism is derived from two Greek nouns meaning "man" and "form." It would be more appropriately applied to the anatomy of animals than to their feelings, which have no obvious shape. Without being disparagingly accused of anthropomorphism, the anatomist can apply to the bones in a bird's wing the same names given to those in a human arm. Indeed, the structural similarities of all terrestrial vertebrates, and the anatomical features they share with fishes, are among the strongest supports of the theory of evolution. It would seem that animals so similar to humans anatomically might also, in some measure, resemble them psychically. It is not evident why anthropomorphism, respectable in comparative anatomy, should be rigidly excluded from comparative psychology. As the minds of nonhuman creatures are more deeply explored and unsuspected capabilities brought to light, anthropomorphism becomes more respectable in biological circles.

The Minds of Birds

Chapter 1

❧

Recognition
of Individuals

A PROBLEM that confronts students of bird behavior is the recognition of individuals. In many species, the members of a pair are so similar in appearance, voice, and behavior that to distinguish them is difficult. The greater the number of individuals included in a study, the more serious the problem becomes. The most frequent method of overcoming the difficulty is to catch the birds in a trap or mist net and burden their legs with narrow bands of various colors in different combinations. On larger birds, numbered wing tags are used. Sometimes one can place spots of paint on their plumage without catching them. Occasionally, missing or rumpled feathers, a trace of albinism, or some other slight abnormality serves to distinguish members of a pair. Len Howard, who for more than a decade kept open house for small birds and lived most intimately with them, learned to recognize individuals, mostly Great Tits, by their facial expressions, characteristic mannerisms, and slight differences in the shapes and shades of their markings. Sometimes, when I have sat long and attentively watching a nest, I have detected distinguishing details in the appearance or behavior of mates that at first looked quite alike.

Our complex human societies would become chaotic without the ability to distinguish individuals (to whom parents give names and governments assign eight- or nine-digit numbers) with diverse functions and relationships. Avian societies are less complex than ours but not so simple that they could persist without the birds' ability to recognize close associates as individuals. The sooner after hatching that birds become mobile, able to wander from their nests and mix with other chicks, or intrude into neighbors' nests placing an added burden on parents not their own, the earlier parent-young recognition develops. Because it is so necessary for their survival, downy, open-eyed precocial chicks, which leave their nests within hours or at most a few days of hatching and follow their parents, learn to recognize particular adults at a surprisingly tender age.

When only two days old, Mallard ducklings could distinguish the voice of their foster parent, a hen, from those of two or three others, as A. O. Ramsay demonstrated by enclosing the hens in boxes and releasing the ducklings at a point equidistant from them. With scarcely any mistakes, each little Mallard toddled to its own invisible mother, guided by her voice. Two-day-old wild Turkeys and Ring-necked Pheasants could likewise distinguish their foster mothers from other calling hens.

When Nicholas Collias mixed domestic chicks of several broods in the dark, they tended to separate, each seeking its calling mother, at least when the hens were of different breeds. In another experiment, Collias placed chicks from mothers of three different colors in an enclosure with three strange broody hens of the same colors. The chicks taken from the black mother followed the black hen, those from the red mother sought the red hen, while those whose mother was white joined the strange white hen. These tests proved that domestic chicks a few days old have some ability to recognize their parents by either voice or appearance. The early attachment of chicks, ducklings, or other young animals to their parents, foster parents, or even to an alarm clock drawn slowly in front of them, is called imprinting.

Although precocial chicks often learn very promptly to recognize their parents, the latter often take longer to recognize their progeny as individuals. Apparently, this is because they may have a large brood, whereas the chicks of most species have only one or two attendant parents, and possibly also because, as in humans, adults have more personality than do infants. Moreover, the appearance of the young changes rapidly, while the parents remain much the same. The changing young often confuse parents, who may be inexperienced birds rearing their first brood. American Coots, attending a brood in which members may differ in age by as much as a week, seem to recognize their offspring by the appearance of the majority. When the oldest of their downy black young begins to acquire lighter plumage, it becomes the odd member of the family, attacked as a stranger by the perplexed parents. Later, when most members of the brood become paler, the youngest, still wearing black natal down, contrasts with the rest and confuses the parents. Sometimes they alternately feed and persecute an odd-colored member of their family, as Gordon Gullion saw.

Until it knows its young individually, a parent of a subprecocial or precocial species, such as coots and gallinaceous birds, will accept and attend others that differ little in age and appearance from its own progeny, although those conspicuously different may be repulsed with pecks if they try to join the family. The more individual attention they give their young, the more parent birds resist the intrusion of strangers. Coots, who feed their young from the bill, repulse with pecks chicks unfamiliar to them; domestic hens, who scratch vigorously to expose food that the chicks pick up for themselves, are hardly more indulgent of intruders; but ducks, who do no more than guide and brood young who find all their own food, are extremely tolerant, often leading a flotilla of ducklings of different sizes and parentage.

In semialtricial birds that breed in crowded colonies on the ground or on cliff ledges, the mutual recognition of parents and offspring when chicks are very young is of great importance. These are mostly marine or fresh-water birds—penguins, murres, gulls, terns, some pelicans—that lay their eggs in shallow nests, often mere scrapes in sand, or even on bare rocks. Soon after they hatch, the chicks can wander over the surrounding sur-face, while the parents bring food to them, often from a great distance, which strictly limits the number of young parents can nourish and makes it imperative that each feeds only its own.

On narrow ledges high above stormy northern seas, Common Murres learn to recognize their parents' voices while they are hatching. In com-pact colonies on dry stream beds in New Zealand, Black-billed Gull chicks could distinguish their parents' voices from those of neighbors when two to four days old, as R. M. Evans demonstrated by playing the parent gulls' recorded voices through a loudspeaker. Because the wider dispersion of Herring Gulls' nests makes it less likely that a chick will wander into a neighbor's territory, the parents need not recognize them so promptly. They accept and feed errant chicks less than about five days old, by which time they can distinguish their own from intruders, whom they now peck.

Sooty Terns, who nest on the ground in compact colonies on small oceanic islands, learn to recognize their own progeny by the fourth day after hatching. In strong contrast to this, Brown Noddy terns, at least where they breed in trees and bushes as on the Dry Tortugas of Florida, where Diane Riska studied them, accept, feed, preen, and brood or shade strange chicks placed in their nests of sticks, even when the aliens differ greatly in age and color from their own chicks. Brown Noddy chicks are not likely to wander from one elevated nest to another. Unless experi-menters interfere, parents are not likely to feed intruding young to the detriment of their own.

In their acceptance of foreign young, Brown Noddies behave like typ-ical altricial birds such as sparrows, thrushes, and warblers. The nestlings of these birds hatch in such an undeveloped state that their parents could hardly recognize them as individuals, and, being sightless, chicks could not recognize their parents except possibly by voice. For some days, the

Common Murres

nestlings will stretch up their gaping mouths for food in response to almost anything that gently shakes their nest or is waved above it; but after their sight develops they crouch down, as though in fear, when something strange, such as a human hand, approaches them. Since the parents know exactly where their nest is situated and the young do not normally leave it before they are feathered and can fly or at least flutter a little, parents treat as their own any hungry, gaping bird they find in it. They may feed and brood well-grown nestlings that an experimenter substitutes for their eggs, or hatchlings that replace older nestlings. Taking advantage of this wide tolerance, parasitic cuckoos and cowbirds foist their eggs on hardworking parents, who often hatch them and rear the ravenous fosterlings at the price of losing some or all of their own progeny. However, birds of a number of species recognize the intruded eggs and cast them out or abandon the nest that contains them. If more birds learned to distinguish alien nestlings from their own, they might be less subject to this imposition.

After young birds leave their nest, the parents often divide the brood between them, the father taking full charge of some, the mother of the others. This division of responsibility and risks promotes efficiency and is practiced by many altricial birds and not a few precocials. By this time parents and young have learned to recognize each other as individuals. Occasionally, a parent will refuse to feed its own begging offspring that is being attended by the other parent, as David Snow recorded of a Eurasian Blackbird, but such rigid division of labor is not invariable. When a hawk killed a male Spotted Antbird in a tropical forest, his mate took charge of the fledgling that he had been attending, in addition to its sibling already in her care, as Edwin Willis noted.

As they grow stronger, the young of certain colonial birds that nest on the ground or on ice or in shallow water, including some pelicans, penguins, and flamingos, gather in pods or crèches, sometimes of hundreds or thousands. Here a few adults guard them while the parents forage at a distance. Returning, each feeds its own young, recognizing it amid the throng. If the parents fed the young at random, the smaller and weaker of them would often go hungry in an undisciplined scramble for meals. The restriction of feeding to a parent's own progeny ensured that a young Greater Flamingo with a broken wing received its food, as L. H. Brown saw at Lake Elmenteita in Kenya.

Birds recognize their mates, as is clear from the fact that most birds repulse strangers of their species that closely approach their nests. A female American Goldfinch incubates with very long sessions, interrupted by recesses too brief to find much food; she depends almost wholly on her mate to nourish her. She can distinguish him, probably by his flight notes, from other males passing by, and when she hears him she twitters charmingly until he alights on the nest and regurgitates thistle seeds to her, as I saw long ago.

A Herring Gull can recognize its flying mate thirty yards away. A gull drowsing on its nest is rarely aroused by the calls of other gulls passing nearby but the voice of its approaching mate arouses it instantly. If both members of a pair of Black-legged Kittiwakes are alive when these little gulls return to their nesting colony after their winter dispersion over the ocean, they recognize one another and reunite, especially if they have successfully nested together in an earlier year. As J. C. Coulson demonstrated by prolonged studies of a colony in England, pairs that have previously bred together tend to have greater success than do newly formed pairs, even when the members of new pairs have already nested with other partners. Prolonged association increases the rapport and cooperation of mates.

An important consequence of individual recognition is the avoidance of inbreeding. An incest taboo appears to prevail in many species of birds. This is especially apparent in cooperative breeders, among which offspring of both sexes remain with their parents and refrain from breeding for one year or longer, helping the parents to defend their territory and to rear further youngsters. When a parent dies, one of the helpers, usually the eldest of the same sex, inherits its position and begins to breed. Thus, among communal-nesting Acorn Woodpeckers, a subordinate remaining with the group avoids the possibility of incest by refraining from breeding as long as its parent of the opposite sex is present; but if this parent is replaced by a woodpecker from another group, the former subordinate may breed. When a breeding male Red-cockaded Woodpecker dies, leaving his mate and her helper son, the latter becomes a breeder and his mother departs. Similarly, Green Woodhoopoes, Florida Scrub Jays, White-fronted Bee-eaters, Arabian Babblers, and other cooperative breeders avoid mating with parents or siblings. Such avoidance implies not only enduring recognition of individuals but likewise continuing awareness of relationships among closely associated birds.

Many birds recognize their neighbors as individuals and may be friendly toward some and hostile to others, as Niko Tinbergen noticed of Herring Gulls. Longtime neighbors, if of good behavior, trust one another more than they trust newcomers. Val Nolan, Jr., followed the interactions of male migratory Prairie Warblers newly arrived on adjoining territories in spring. Some had been settled on adjacent territories in the preceding year, others had no known earlier association. Among thirty-three pairs of former neighbors, he saw only two fights, both in unusual circumstances; among seventy-three pairs of males who had not previously been neighbors, he witnessed nineteen territorial skirmishes. Clashes were four times as frequent between new neighbors as between old neighbors, who evidently remembered each other from the preceding summer.

Among birds in flocks, a dominance hierarchy or rank order frequently develops. Domestic hens reveal their social status by occasionally pecking their subordinates. A can peck all the others, B all except A, C all but A and B, down to the lowest-ranking, X, which may be pecked by all but does not ordinarily retaliate. A new hen entering the flock ranks near the bottom but, if of a vigorous personality, may work her way upward. A hen who becomes sick or is injured falls lower in the hierarchy but, if she recovers, may regain her higher status. The situation among hens is known as peck right. Domestic pigeons behave rather differently. The pecked subordinate does not always submit passively to this abuse but may on occasion return the insult. However, A strikes B more frequently than it is struck by B, a relationship that is known as peck dominance.

Similar rank orders develop among birds attending feeders and probably more rarely among birds flocking in forests and fields. With incredible patience, Winifred Sabine disentangled the social relations of two flocks of Oregon, or Dark-eyed, Juncos on the West Coast. In one assemblage, she learned the order of dominance of forty-two birds. She could recognize individuals by the color bands she had attached to their legs, but they probably did not depend upon these artifacts. It would hardly be possible to develop and maintain such a scaled hierarchy unless each junco recognized individually most, if not all, of its forty-one associates in the flock.

Birds recognize other members of their species by voice, appearance, and comportment. As with us, the face or head seems to be of first importance for individual recognition. By molting, some birds change their colors as much as we often do when we change our clothes, but facial expression is more constant. Hens whose combs were fastened on the wrong side of their heads, enlarged, or otherwise altered by A. M. Guhl and M. M. Ortman were treated as strangers by their flock mates. Artificially coloring a hen's head or neck evoked much stronger reactions by her flock mates than dyeing her body or changing her size by attaching or removing feathers. Years ago, when I was just beginning to study birds, I whitened the crown of a Rufous-breasted Castlebuilder, or Spinetail, to distinguish it from its mate, who did not differ from it in plumage. The poor bird was rejected by its partner, and I have never again made such a drastic alteration in a bird's appearance.

Birds are hardly less acute at recognizing humans than they are at recognizing other birds. They appear to recognize human individuals better than we can distinguish them. In *King Solomon's Ring,* Konrad Lorenz told how, one Saturday afternoon at the Altenburg railroad station in Austria, he descended from a train in a crowd of holidaymakers and noticed his tame Yellow-crested Cockatoo flying high overhead, in danger of becoming lost. Not knowing what else to do, he bellowed out at the top of his voice *o-ah,* his imitation of the bird's call. After brief hesitation, the cockatoo folded its wide wings and dived down to alight on the distinguished ethologist's outstretched arm, singling him out amid the throng.

Dana Gardner, the artist who has illustrated many of my books, told me of a similar episode with his pet Feral Pigeon, Crystal. One day Dana was marching, a mile from home, among fifty members of his high-school band, with a large tuba encircling his body and projecting above his head, when Crystal happened to fly by. Recognizing Dana in the crowd despite the instrument that partly concealed him, the pigeon spiraled out of the sky and landed near his feet. Then, to the amazement of the other musicians, Crystal continued for several minutes to march beside his master.

While Derek Goodwin was stationed at an army camp in Egypt during the Second World War, he often amused himself by gathering scraps of food after meals and giving them to the Black Kites that frequented the camp. Although these scavengers were, with good reason, highly suspicious of people, one of them quickly learned to recognize Goodwin as a benefactor. After he had fed it only twice, he was walking from the dining hall to his tent, by a path that others in identical uniforms were constantly using, when he noticed the kite perching on a pole. Having stared intently at Goodwin for a few moments, the black bird flew straight to him and hovered overhead to seize the tidbits that the soldier flung into the air for it. Thereafter, this kite frequently approached Goodwin, especially when an upward glance revealed his face.

To learn how birds recognize people, observers sometimes try to deceive them by unfamiliar attire. The German ornithologist Oskar Heinroth had a pheasant who courted him and fought his wife as a rival. When the couple exchanged clothes, the bird started to attack Herr Heinroth in his wife's dress, stopped, scrutinized their faces, then flew at Frau Heinroth in trousers. When Katarina Heinroth and her sister exchanged garments, the pheasant still distinguished his "enemy."

Althea Sherman wrote about a Red-winged Blackbird named Brigham, who developed an antipathy to her because she frequently visited one of his several mates' nests. Not only did he batter her in the Iowa marsh where he dwelt, he would advance to her dooryard to shout defiance with flashing red epaulets. He did not attack other people who approached his nests. When Sherman and her sister visited the marsh dressed in each other's clothes, Brigham was confused; after an interval he struck the sister feebly a few times, without touching Althea. Apparently, the bird recognized his "enemy" mainly by her garments but did not disregard her face. The discrepancy between clothes and countenance softened Brigham's aggression.

A pair of American Robins, whose young Constance Nice weighed daily, greeted her with loud outcries whatever she wore but ignored all other people, according to her mother, M. M. Nice. Some years ago, I had a similar experience with the robin's tropical counterpart, the Garden Thrush, also called the Clay-colored Robin. For two years, an exceptionally nervous and querulous female nested in our Costa Rican garden. During the first year, I incurred her disapproval by occasional inspections of her nest and by gathering oranges from a nearby tree. She never attacked me but protested with her plaintive *quee-oo* whenever she saw me, although mostly she ignored other people. A year later, she remembered me and displayed her special antipathy even before I found her nest in a more secluded spot. Whenever she saw me approaching, she would fly far to meet and follow me over several acres, incessantly complaining. She would raise her mournful cries even when, through an open window, she glimpsed me inside the house, although most birds pay little attention to what is happening indoors.

To test the thrush's ability to recognize me, I walked into the garden in a fantastic garb such as I had never before worn. These appearances were sometimes greeted by a few complaints but not by the close following and reiterated cries that I usually evoked. When I was fairly confident that the thrush failed to recognize me in my disguise, I decided to see how she would react to a sudden change in my appearance. After walking a short way in my everyday clothes, with the bird following and protesting loudly, I donned in her presence the dressing gown and felt hat that I had carried inconspicuously under an arm. She repeated *quee-oo* a few times more,

Garden Thrush

then fell silent and ceased to pursue me. The removal of this extraordinary costume evoked a fresh outburst of complaints. I concluded that the excitable Garden Thrush recognized me by my total appearance rather than by my face alone, but sometimes she seemed to suspect my identity beneath a strange attire. A diligent search through the vast literature about birds would probably yield enough instances of birds' recognition of individual humans to fill a chapter, if not a book, but the foregoing are enough to indicate that a great diversity of species, from raptors to songbirds, have this ability in no small measure.

I have started this book with an examination of birds' ability to recognize individuals, from their own family members to humans, not only because it reveals the keenness of their vision and their minds but also because this ability is indispensable for the preservation of family unity or any social organization beyond the most rudimentary. What would our human societies become if we did not recognize individuals and establish special relations between them but treated alike all of the same sex and age? Avian societies are also established upon the ability to recognize individuals. Monogamy, practiced by about 90 percent of the species of birds, would be impossible if mates did not know each other intimately and respond to each other in special ways. In cooperatively breeding groups, widespread among birds permanently resident in mild climates, each member knows all the others closely and responds to them in diverse ways. Even larger gatherings of foraging birds, as at feeders, might become less orderly if the attendants did not recognize one another individually and establish a *modus vivendi* among them. Recognition of individuals is proof that bird's minds are capable of fine discriminations and establishing order in complex situations.

References: Bent et al. 1968; Brown 1958; Collias 1952; Coulson 1972; Evans 1970; Goodwin 1978; Guhl and Ortman 1953; Gullion 1954; Howard 1952, 1956; Lorenz 1952; Nice 1943; Nolan 1978; Ramsay 1951; Riska 1984; Sabine 1959; Sherman 1952; Skutch 1976; Snow 1958; Stacey and Koenig, 1990; Tinbergen 1953, Willis 1972.

Chapter 2

Memory and Anticipation

BIRDS ARE commonly supposed to have short memory. They seem promptly to forget a narrow escape from capture by a hawk or the loss of eggs or young to a nest robber; surprisingly soon after such an apparently distressing experience, they may resume singing. However, song is not an unfailing indication that a bird has forgotten its fright or its loss; strong emotion of any kind may elicit song. Birds are aware of more than immediately present stimuli; they remember the past and anticipate the future.

While I occupied a cabin in a narrow clearing in a Panamanian forest, a pair of Crimson-backed Tanagers nested in a nearby shrub. One afternoon, when their nestlings were becoming feathered, the parents complained so persistently, drawing a crowd of feathered onlookers, that I went to investigate. A careful search through the surrounding herbage and bushes failed to reveal a cause for their disquiet. I returned to my writing in the cabin, but the tanagers' continuing cries distracted me from my notes.

After several searches failed to reveal a reason for the birds' anxiety, I watched from a blind. The tanagers repeatedly perched on a low branch of a rosebush, cocked their heads to one side, and peered down into the weeds two feet below, at the same time reiterating their full, nasal cries. With this information, I found what troubled them: a large, black-and-yellow Mica, a snake that preys insatiably on eggs and nestlings. I dispatched the snake while the birds remained at a distance. Returning, they continued to peer at intervals into the weeds, to see whether their enemy still lurked there. This pair of tanagers, who had recently lost a nestling to each of two snakes, were so agitated by their long interval of tension that for the next hour they neglected to feed their nestlings, who were calling

Crimson-backed Tanager

for food. They did not promptly forget the peril to which their brood had been exposed.

Another indication that birds remember their fears and losses is their persistence in bringing food for vanished nestlings. For hours or days, I have watched bereaved parents continue to take food to nests desolated by predation, or even to the sites of nests that have been carried off by nest robbers. I have witnessed such behavior by birds as diverse as trogons, nunbirds, wrens, vireos, and caciques. For six days, a pair of Golden-naped Woodpeckers brought offerings to their desolated nest cavity. Others have reported similar behavior by hummingbirds, kingfishers, and finches. Murres, or guillemots, ceremoniously offer food on seaside cliffs from which their pipped egg or chick was lost two or three days earlier. A captive Wood Thrush carried food for eight days after the death of newly hatched nestlings. In altricial birds, the usual stimulus for feeding is the gaping mouths or hungry cries of nestlings. In the prolonged absence of an appropriate stimulus, do they persist in this activity from blind habit or because they remember their lost offspring? Apparently, they can hardly accept the fact that the brood to which they have devoted so much effort has suddenly been snatched from them.

Birds' spatial memory, their ability to relocate places or points in space, is impressive. After a Rufous-tailed Hummingbird's nest collapsed, I fashioned a shallow cone of stiff paper, lined it with cotton, tied it as close to the fallen nest's site as I could, and placed the two nestlings in it. Returning to feed them, the mother hovered at the exact site of the vanished nest before she dropped to the artificial nest a few inches lower. Even after she had visited the nestlings several times in their new abode, she continued to hover first before their old position, neglecting for the moment the artificial nest in full view. A human finding a nest in the midst of a forest or in a wide meadow may have great difficulty relocating it without having carefully studied its situation in relation to nearby landmarks, such as a readily recognized tree, or having set a stake beside it in the meadow. Yet without special guideposts, birds unerringly find their nests amid a rather uniform expanse of woodland, thicket, or meadow, or in the midst of thousands of other nests in a crowded colony.

When food is abundant, some birds cache or store it for future recovery. Such forehandedness is widespread among jays, other members of the crow family, and titmice. In autumn, Pinyon Jays of western North America store many thousands of seeds, but the greatest of all cache makers are the nutcrackers. Also in western North America, Clark's Nutcrackers store seeds of pines and other conifers. Widespread Eurasian Nutcrackers cache hazelnuts in regions where wingless pine seeds are unavailable. For distances up to ten miles (15 kilometers) or more, a Clark's Nutcracker transports a mass of seeds in its sublingual pouch, a distensible pocket opening beneath the tongue in the floor of the mouth. When it reaches its caching area in the woods or on a windswept ridge where snow will lie less deeply, the bird loosens the soil with its sharp bill, pushes in one or more seeds, and covers them with soil or ground litter. Then it distributes others in neighboring spots. Individual nutcrackers have stored from 20,000 to an estimated 100,000 seeds in autumn when they ripen. The caches of several flock members may be intermixed in the same storage area, but each individual apparently harvests its own. These stores will nourish the industrious birds through the severe winters of high latitudes and altitudes, and help feed their nestlings and fledglings early in the following spring.

How nutcrackers, Pinyon Jays, and other jays retrieve their seeds is a problem that has been studied by many investigators, in the field and in aviaries, with results that have been reviewed by Stephen Vander Wall in *Food Hoarding in Animals.* In natural and experimental situations, recovery rates of stored seeds have ranged from about 50 to 99 percent, or from about half to nearly all the birds' attempts to find their buried seeds by digging into the ground have been successful—a rate too high to be attributed to chance. Since the signs of soil disturbance made when the seeds were buried are soon obliterated by weather, they cannot after months guide a bird to its cache. Olfactory cues appear to be ruled out by the fact that nutcrackers can retrieve their seeds through a thick blanket of snow—up to thirty-three inches (85 cm) in the Eurasian species, as they must do to stay alive through long, snowy winters.

The birds rarely find food hidden by others. Evidently, each time they

Clark's Nutcracker

bury seeds in a hole, they pay careful attention to its exact position in relation to nearby landmarks: trees, stumps, rocks, and the like. They prefer
for orientation vertical objects that will project above snow to horizontal
ones that might be covered. When these signposts are experimentally
shifted, the birds dig for their buried nuts in the correct spots relative to
the markers but, of course, fail to retrieve their food. When the buried
seeds are removed while the landmarks remain in their original positions,
the birds dig where they had buried food, thereby providing additional
evidence that they do not rely upon olfactory cues. To achieve their high
rates of retrieval, each bird must remember hundreds or thousands of
cache sites with almost photographic accuracy. Without an astonishing
spatial memory, they might recover few of their laboriously stored seeds.
Recent experiments by Russell Balda and W. Wiltschko have disclosed
that Scrub Jays use the Sun as a compass to aid in the recovery of their
buried seeds. In the absence of direct sunshine, they are apparently guided
by polarized light from the sky.

Whereas sedentary jays and nutcrackers demonstrate their spatial memory in a small area, long-distance migrants display a similar capacity in a
much vaster theater. After their long flight from their breeding grounds
in Canada and the northern United States, a number of Northern Waterthrushes settled in the large botanic garden in the center of Caracas,
Venezuela. Here they were banded by Paul Schwartz, who discovered that
individuals occupied the very same territories year after year, remaining

for about six months—longer than they stayed on their nesting territories in the north. When he captured a number of these waterthrushes and released them at distances of six to forty miles (10 to 65 km), four birds who had occupied their territories in the botanic garden for more than one year returned to them, but fourteen new visitors to the garden failed to find their way home. Innately equipped with the skill to reach the broad South American continent from points three or four thousand miles away, they had settled in the botanic garden without previous knowledge of it. To return to its winter territory, a waterthrush must learn and remember details of Venezuelan topography as well as its precise position in relation to surrounding objects. Apparently, on their northward migrations they learn the route well enough to retrace it in the opposite direction five or six months later—no mean feat of learning and memory!

In their winter homes in Middle and South America, none of the many migrants from farther north is known to nest. Some pass the months of the northern winter in wandering flocks but others promptly establish individual territories where, if all goes well, they will remain until the following spring. That an individual returns to the same territory in successive years has been proved by observations of banded or otherwise recognizable birds. In the state of Veracruz, Mexico, John Rappole and Dwain Warner found Wood Thrushes, White-eyed Vireos, Kentucky Warblers, Hooded Warblers, and six other species occupying the same territories in more than one year. Such site fidelity appears to be widespread among migrants with individual winter territories. Although the Yellow-bellied Flycatcher who catches insects in our Costa Rican garden is not marked, from its behavior I surmise that the same individual remembers its way to this mountain-rimmed valley and honors us with its presence year after year.

Many migrants appear to commute annually between definite, remembered summer and winter territories that are often thousands of miles apart, perhaps in a garden in Colombia or Venezuela and in a clump of spruce trees in Canada. Of 75 cases in which male Prairie Warblers returned to Val Nolan's study area in Indiana, after having bred there previously, in 55 (73 percent) they reoccupied their former territories. Three males held the same territory for five years and one for six years. Females were much less strongly attached to their former territories; in only 14 of 137 cases (10 percent) did they nest in the same places in successive years. Female migrants reoccupy a breeding territory less often than males do because, returning later than the latter, they often find that an earlier arrival of their sex has joined their former partner, and they must look elsewhere for a territory and a mate.

The Common Tern is one of the most cosmopolitam of birds, nesting over much of the northern hemisphere and, after they breed, spreading widely over all the oceans. Of the many terns J. C. Welty reported banded by Oliver Austin on Cape Cod, Massachusetts, 2,964 returned to their natal colony, and 76.5 percent of these birds chose the same nest sites in consecutive years. Some returned to the colony three, four, five, or more times, and as the years passed their site fidelity increased. To locate a known nest site from afar requires not only supremely competent navigation but also retentive memory of local topography.

Racing pigeons, on being released, return to an old loft from which they have been absent for years. Domestic hens recognize a former home after an absence of several years. Birds retain strong memories of places where they have been threatened or alarmed. Margaret Nice noted that a tame Song Sparrow was deeply agitated when taken into a room where he had been shown a stuffed Barred Owl more than four months earlier. The same sparrow was even moderately alarmed when exposed to the outline of a model of an owl that had frightened him thirteen months earlier and that he had not seen in the interval. Retentive memory of places and situations that have frightened birds has great survival value.

Birds that sing only in their nesting season repeat the same songs year after year, remembering them during the long months when they are latent. An amazing capacity to learn and remember the songs of other birds is exhibited by male Marsh Warblers. When, a few weeks after they hatch in central Europe, they leave for their winter quarters in Africa, they have heard and stored in memory many songs of their neighbors, which they do not include in their juvenile chatter. On their southward migration through eastern Africa, they pick up more alien songs, and while wintering in southeastern Africa they continue to enrich their repertoire of borrowed motifs. During their sojourn in Zambia, from late December to early April, they sing with increasing frequency until their northward departure. Few wintering passerine birds are so songful.

Before arriving at the age of ten or eleven months, on the marshes where they nest Marsh Warblers appear to have lost their juvenile aptitude for learning new songs; but they have already amassed an extraordinarily rich repertoire. An individual warbler can now imitate from 63 to 84 species, an average of 76. By comparing sonograms of the warbler's songs with those of African birds, François Dowsett-Lemaire identified 212 borrowed motifs in the songs of twenty-nine Marsh Warblers. This number included the voices of 80 African passerines and 33 nonpasserine birds, a total of 113, plus 99 European species. A remarkable aspect of this performance is that all the European songs were learned in the warblers' homeland while they were still immature. Perhaps the birds repeated these songs in Africa but they were not voiced in Europe until seven or eight months after they were learned there. By the time yearlings begin to breed, they have established a song pattern, largely a medley of imitated notes, which they will continue to use in subsequent years. With such a wide variety of phrases gathered on two continents, Marsh Warblers can sing uninterruptedly for an hour without becoming monotonous. Few birds chant so continuously. Their unique solos attract mates and appear to have been promoted by female preferences, or sexual selection, which can favor brilliant singing no less than gorgeous plumage.

Birds may remember songs and places better than they recall other individuals of their kind. How long domestic fowls remember one another depends upon how long they have previously been together. Roosters or hens reintroduced after a separation of only two weeks may treat each other as strangers. Black-crowned Night-Herons remember their associates for at least twenty-two days, White-throated Sparrows for seventeen days.

The memory of free birds is difficult to assess. Many, especially permanent residents and geese and swans that migrate in family groups, retain the same mates as long as both live; but since the partners are almost constantly together or join one another at short intervals, their familiarity is continually renewed, and we lack proof that they would remember each other after a long separation. Migratory or seasonally mated birds are often found with the same partners after being apart for more than half a year; but we do not know whether individual attraction or site fidelity—the attraction of a territory or nest site where they have bred successfully in a past year—brings them together again. Remembering the mate and remembering the territory may act together to reunite the partners. Black-legged Kittiwakes—who, as told in the preceding chapter, breed more successfully with well-tried partners—evidently recognize those partners as individuals after three to eight months of wandering apart over the ocean. Likewise, the migratory Prairie Warblers who dwelt more peacefully with former neighbors than with new neighbors remembered them after six or seven months' absence from their breeding grounds.

Not only do birds recognize individual humans, even amid crowds, but they remember people for a long time. Len Howard, who without artificial aids knew her birds individually, found that they clearly remembered her. A Great Tit, readily identified by a crooked leg that had been fractured, flew straight to her hand after she had been absent for two years. In a wood half a mile from her garden, another Great Tit, Pippa, flew directly to her, trailing four fledglings, and fed them with food from the woman's hands. They had not met for fifteen months.

A Eurasian Robin who had nested in Howard's garden flew to her while she stood talking to someone, at a distance from the garden, eight months after she had last seen the bird. He ate greedily of the food she al-

Prairie Warbler

ways carried for birds she happened to meet on her walks. The robin would not approach other people who tried to coax him to their hands. A continent away, Althea Sherman was certain that one of the Ruby-throated Hummingbirds that drank syrup from bottles she provided for them recognized her after almost eleven months. As already told, the Garden Thrush who disliked me remembered me from one breeding season to the next, but she may have seen me in the interval.

Memory and anticipation have a common foundation. Both are rooted in the mind's ability to represent what is not immediately present—what actually happened in the past, what is expected in the future. We wonder whether a bird starting a nest foresees its finished form or anticipates the eggs she will lay in it, the nestlings that, if all goes well, she will feed and brood there. We cannot prove that they do, but the behavior of certain birds, especially males, that I have called anticipatory food-bringing, is suggestive. Days before the date of hatching, they bring food to the nest, peer in intently, often with soft notes as though to arouse drowsing nestlings, and present the food to the eggs. That this behavior differs in motivation from feeding the incubating female, as many male birds do, is evident from the fact that even males who do not feed their incubating mates on the nest offer food to the eggs when the mate is absent. If she happens to be sitting when he arrives, she may accept what was not intended for her, pass it back to him, swallow it tardily, or refuse it. Anticipatory food-bringing has been recorded of several kinds of wood warblers, tanagers, finches, the Tropical Pewee among flycatchers, and the Masked Tityra. A male Variable Seedeater even offered to regurgitate food at a second-brood nest in which his mate had not yet laid an egg. Female birds, whose closer association with nest and eggs keeps them better informed of the situation there, have more rarely been seen to bring food before their eggs hatch; but a Fasciated Antshrike, Tropical Pewee, Dusky-capped Flycatcher, and Orange-billed Sparrow did this while I watched.

Surprisingly, many female birds have no call or other signal to apprise a mate not sharing incubation that the progeny have hatched and the male is needed to help feed them. The male must discover for himself that he now has paternal obligations. Among behaviors that help him learn that his nestlings have hatched are occasional visits of inspection to the nest (without food) and anticipatory food-bringing. Prompt feeding by the male is of special importance when it enables the mother to remain more constantly on the nest on cold or stormy days when hatchlings need much warming. Nevertheless, we must ask why males present food to eggs that do not provide the usual stimulus for feeding nestlings, their gaping mouths and begging cries. Nolan, who carefully investigated this problem with Prairie Warblers, regarded anticipatory food-bringing as a "vacuum activity." In the absence of conditions appropriate for a certain behavior, the energy or the impulse to engage in it sometimes intensifies until it is performed gratuitously, as though in a vacuum, as when a bird indulges in nest-shaping movements without a nest site, or a person who has long been solitary begins to speak aloud.

In the mountains of Guatemala, many years ago, a male Pink-headed Warbler puzzled me by trying earnestly, five times in a day, to deliver food to eggs a week from hatching. Finally, it dawned on me that he was anticipating his offspring, behavior of which I had never heard. I surmised that he was impatient to feed nestlings because he remembered them from a former brood, but without knowing his history, I lacked evidence for this. Among the thirteen Prairie Warblers that Nolan and his assistants watched bring food to eggs thirty-four times were a few yearlings who probably had no previous experience of nestlings. This does not exclude the possibility that they had an innate image, or racial memory, of nestlings, as birds apparently have of the nests they build. As in many other situations, we cannot be sure of what is happening in the mind of a bird or any other animal, but I think it probable that the parent offering food to young still tightly enclosed in the eggshells foresees his or her nestlings.

Birds, especially females, are widely believed by ornithologists to choose breeding territories with some regard for their quality as sources of food for their future nestlings, as well as for the male territory holder who will help rear young. Migratory birds often select their territories early in spring, before the vegetation is well grown and productivity has approached its maximum, as it should do weeks or months later when nestlings will need much nourishment. A realistic forecast must be based on individual experience or racial memory encoded in the bird's genes. Again, we do not know how the prospecting female assesses her breeding territory, but her choice appears to involve complex mental activity, including foresight.

The foregoing observations should make us reconsider the widely held view that birds live only in the present, without conscious memories of the past or anticipation of the future.

References: Balda and Wiltschko 1991; Coulson 1972; Dowsett-Lemaire 1979; Howard 1952; Nice 1943; Nolan 1958, 1978; Rappole and Warner 1980; Schwartz 1963, 1964; Sherman 1952; Skutch 1953, 1954, 1976; Wall 1990; Welty 1975.

Chapter 3

❀

Social Life

LIFE IN complex societies tends to develop subtle relations between individuals, to sharpen intelligence, to promote communication, and to increase awareness of self in relation to others. Of all nonhuman animals, birds, especially the smaller kinds, take the most thorough care of their young, building for them nests that are often elaborate, feeding them, warming them, cleaning them, defending them, in many species leading them to sleep in nests that they have just left or in others specially prepared for them. At low latitudes, parental care tends to be more prolonged than at high latitudes, where breeding seasons are shorter and migration, or wandering in the inclement months, usually dissolves families. Sometimes the young of permanently resident birds, fully grown and well able to forage for themselves, remain with their parents for a year or more, helping them to rear subsequent broods in cooperatively breeding groups, the most advanced societies among vertebrate animals, apart from humans. Over one hundred cooperatively breeding species are now well known, and the number increases as ornithologists extend their researches into less accessible regions. No attempt to fathom the minds of birds would be complete without consideration of the significance of cooperatively breeding associations.

These associations are of three main types. The most frequent consist of a single breeding pair with helpers who are mostly their progeny. Less frequently, the group contains two or more pairs breeding in as many nests, with their assistants, all mutually supportive. In the third system, two or more females lay their eggs in the same nest, and all members of the group help to rear the enlarged brood, without respect to parentage. The boundaries between these systems are not always sharp; combinations of them are known. Jerram Brown designated the first system singular breeding; the second, plural breeding; and the third, joint nesting.

Cooperative breeding is also surveyed in a large book edited by P. B. Stacey and W. D. Koenig, and in my *Helpers at Birds' Nests*.

A member of the Old World babbler family (Timaliidae), the Gray-crowned Babbler of eastern Australia offers an example of singular breeding. It was intensively studied for three years by Brian King in southern Queensland. This ten-inch (25-cm) bird has a fairly long, sharp, curved bill and strong legs. Above, both sexes are grayish brown, with white-tipped tail feathers. They have white eyebrows, dusky faces, white throats and chests, and pale brown abdomens. The color of the iris changes slowly from dark brown in juveniles to yellow in mature birds more than two and a half years old, thereby revealing to group members the age and status of individuals, which appears to be of importance to cooperative breeders. In open woodland with herbaceous ground cover, these babblers live in stable groups of two to thirteen birds, with equal numbers of males and females. Almost wholly insectivorous, they forage on the ground and over trunks and branches of trees but rarely amid foliage. On the ground they overturn large objects by inserting their bills beneath them and pushing forward and upward. On trees and logs they pry off loose flakes of bark and insert their bills deeply into crevices and logs to extract prey.

As in other cooperatively breeding birds, all group members dwell in close, mutually helpful amity. Often one offers food to another, regardless of age and in the absence of begging. The recipient may pass it to another bird, who in turn may give it to a third. In late afternoon, before going to rest, the babblers dust-bathe close together, flicking dry earth and sand over themselves and their companions with their bills while they fluff out their plumage and flutter their wings. Then they fly up to a tree and diligently preen one another as well as themselves, sometimes three birds attending to a fourth. Whether arranging their feathers or drowsing on a perch, they rest in close contact. At night the whole group, except the female incubating her eggs or brooding her nestlings, crowds into a roofed dormitory nest, of which several are always available in each territory and are kept in good repair by all group members. Unlike many other birds, these babblers apparently never sleep in old brood nests.

Groups of Gray-crowned Babblers defend territories of eleven to thirty-seven acres (4.5 to 15 hectares) or forage over much larger, undefended home ranges. In each group the single breeding male is dominant; his mate, the single breeding female, is next in the hierarchy, with the others ranked in the order of their ages, the youngest lowest. They rarely quarrel. When a member of the breeding pair disappears, it may be replaced by the advancement of a nonbreeding adult or subadult of the same group

Gray-crowned Babblers

or by an immigrant from another group, which has the advantage of avoiding inbreeding.

In cooperative breeders, group displays are frequent. The huddle display of Gray-crowned Babblers begins when the breeding pair fly together to a branch and rapidly repeat chuckling or barking notes. Hearing this signal, other group members join them and call with the same sounds. Bunched together, they posture with bodies held low, tails widely spread, and partly extended, fluttering wings. Presently the mated pair stand upright amid the chuckling crowd and call antiphonally. When a huddle occurs while the primary female is in her nest, one or more of the nonbreeding females will join the male in antiphonal calling. These displays are given in all seasons and especially when a trespassing group or lone babbler is encountered in the home territory, as though to demonstrate or reinforce the resident group's solidarity and determination to defend its domain. In boundary disputes, each group huddles on an exposed branch in view of the other, sometimes continuing for five minutes. Between repeated huddles, members of the opposing groups chase each other but rarely fight, as is true of most cooperative breeders in similar situations. While these confrontations continue, members of one group may feed begging young of the rival group, as King saw eleven times.

When a babbler is caught by a predator or held in a human hand, its cries immediately bring every other babbler within hearing, of whatever group, to join in simultaneous distraction displays around the distressed bird.

The Gray-crowned Babblers' breeding nest is a bulky, closed structure, entered through a short tunnel under an overhanging projection. Situated among the outer branches of a tree or shrub, five to fifty feet (1.5 to 15 m) above the ground, it has an external diameter of twelve to twenty inches (30 to 51 cm) and encloses a chamber about eight inches (20 cm) in diameter. Within a frame of coarse sticks is an inner wall of fine twigs and grasses, lined with grass, feathers, thistledown, fibrous bark, and other soft materials. Like the dormitory nests, these brood nests are built by all group members of both sexes. In such a structure, the single breeding female of most groups lays two or three eggs. Rarely a larger set, containing eggs differing in size, shape, and color, suggests that they were contributed by two or three females, members of as many pairs rather than consorts of a single polygynous male.

During the long incubation period of twenty-one to twenty-five days, only the female sits, for intervals of about twenty-five minutes, rarely as long as an hour. At two nests, she was present for 67 and 68 percent of her active day. While she covers the eggs, all group members visit her, bringing nest materials or gifts of food with increasing frequency as the date of hatching approaches. During the twenty to twenty-two days that the young are in the nest, all members of her group feed them. At one nest, three nonbreeding helpers contributed more meals than the two parents together, but at all nests the father was the most active provider. After the young fledge, they are nearly always led to sleep in a dormitory nest with the older group members, all of whom respond to their pleas for nourishment. Now, in some groups, the mothers feed the young substantially more than do the fathers. Older group members feed both nestlings and fledglings more often than younger ones do.

Only one of King's nests yielded fledglings. In a different study of Gray-crowned Babblers in southern Queensland, James Counsilman determined that their breeding success increased with the age of mated pairs and the number of their experienced helpers. The ratio of fledglings and juveniles to older individuals was greater in larger groups. Those with only two birds more than one year old had an average of only 0.4 younger members; with three experienced individuals, 1.3 young; with four, 1.8; and with six, 3.7 younger members. The more helpers, the more productively Gray-crowned Babblers nest, but this is not true of all cooperative breeders.

For an example of plural breeding, we turn to the Chestnut-bellied Starling of Africa, which Roger Wilkinson studied for three years on and around the wide campus of Bayero University at Kano, Nigeria. Here, in savanna country, the starlings nested among the gardens and introduced trees around the university buildings and on the farms surrounding the campus. These starlings live in social groups of nine to thirty-five individuals, each group containing from two to six breeding pairs, nonbreed-

Chestnut-bellied Starlings

ing adults of both sexes, and juveniles. Alike in their dark plumage, adult males and females can be distinguished only by behavior. Fledglings lack the glossy chestnut breasts of adults and have dark eyes and yellow bills instead of the cream or white eyes and black bills of older birds. Although when about six months old they resemble adults in bill and eye color, males do not breed until two or sometimes three years old. At Kano, males outnumbered females by approximately 1.3 to 1. As in other cooperatively breeding birds, they usually remained to breed in their natal group, whereas young females more often migrated to neighboring groups.

In this region of two annual wet and dry seasons, the starlings have two breeding seasons each year, one before and one after the main rains, which fall from July to September. In a single group, two to six pairs nest simultaneously. The male and female of a breeding pair share rather equally the work of building, chiefly with dry grasses, a dome-shaped nest with a side entrance, in a low, thorny tree, high in a clump of mistletoe, or upon an abandoned White-billed Buffalo Weaver nest of sticks. While the female incubates alone, her mate rarely feeds her in the nest.

After the nestlings hatch, helpers begin to feed them and to remove their droppings. Although they rarely participate in building a nest, these assistants occasionally bring grasses or feathers to it while it holds young. As the nestlings grow older, the number of their attendants increases, until before fledging as many as fourteen are sometimes active at one nest, although seven is more usual and a few nests lack helpers. As early as six weeks after they fledge, juvenile starlings begin to feed nestlings, but they tend to restrict their visits to later nests of their own parents. Yearlings spread their benefactions more widely. The most active feeders are birds in their second year, who help, simultaneously or successively, at nearly all the nests of different pairs available to them in their group. Adult non-breeding males distribute their aid among the several nests about as widely. Breeding males helped at nearly half the nests at Kano. Breeding females tended to nourish only their own nestlings, but occasionally they took contributions to neighbors' young. Nonbreeding females, who often migrated to a different group, distributed food among the nests only slightly more widely than breeding females did. Rarely a starling took food to a nest in a group adjacent to her own, and she might have done this more often if the resident birds had not repulsed her. After the young left their nests, helpers as well as parents continued to attend them.

At Kano, nests with seven to fifteen attendants produced significantly more fledglings than did nests with two to six attendants, but it was not clear whether this was because the former had more helpers or more eggs. Even so, the advantage of having helpers was evident: nests with seven or more attendants fledged 82 percent of the nestlings that hatched, whereas those with six or fewer attendants fledged only 63 percent. Although the Chestnut-bellied Starlings' breeding system, with active nests of several females served simultaneously or successively by a number of group members, is less usual among cooperative breeders than is a single reproductive pair whose nest receives all the attention of group members, it nevertheless appears to be a highly successful arrangement.

Another starling that follows a similar system is the Superb of East Africa, with groups of up to twenty individuals, including several females who lay and incubate their eggs in separate nests. From mid-October to the end of the following June, six breeding females of one group nested at least twenty-two times. At the nine successful nests of this clan, the average number of attendants feeding nestlings was 12.2. Only a month after they leave their nests, juveniles help at other nests.

In the New World, at least five species of jays defend group territories in which several pairs breed simultaneously. Among the largest of these groups of plural-nesting jays are those of the southern race of the blue-and-black San Blas Jay. John William Hardy and his associates studied these birds intensively during five breeding seasons, in groves of coconut palms and neighboring patches of native woodland, near Acapulco in the Mexican state of Guerrero. Here groups of thirteen to twenty-six birds a year old or older resided permanently. Each group contained from six to ten breeding pairs that retained the same partners from year to year, plus a smaller number of nonbreeders. Only one female laid within each nest of interlaced sticks in the crown of a coconut palm, and during the seventeen or eighteen days that she incubated her three or four eggs she was fed only by her consort.

San Blas Jays (immature on left)

After the nestlings hatched, helpers began to attend them. Except incubating or brooding females, most breeders assisted simultaneously at two or more nests in their group, and some brought food to every nest with young. While building their own nests, these jays fed the occupants of more advanced nests. In addition to feeding their incubating mates, males found time to assist other pairs with nestlings. Breeding females served as helpers before their eggs were laid and after their young fledged or were lost. However, while feeding their own nestlings, parents seldom visited other nests. Yearlings and other nonbreeders spread their attentions just as widely. These jays helped their parents when the latter had dependent young, but they began to assist at the first nest that produced nestlings, regardless of relationship. Some broods were nourished by as many as fourteen individuals, including the parents, neighboring breeders, and younger birds.

When about eighteen to twenty days old and hardly able to fly, the young jays left their nests and fluttered to the ground, where their attendants led them to cover. San Blas Jays were careful not to cross the boundaries of their home ranges and avoided confrontations with neighboring groups. Even when members of Hardy's team carried fledglings, in full view of their attendants, into the home range of an adjoining group, the older birds refused to follow their calling young into foreign territory. However, the resident jays promptly adopted alien fledglings left among them, caring for the newcomers as though these were their own. Most young and adult jays stayed in their natal group from year to year; the few that emigrated were mostly females. The annual survival rate of yearlings and older birds was in most years about 75 percent, which is much better than that of jays at higher latitudes. Even among cooperative breeders, San Blas Jays are outstanding in the number of individuals that interact in friendly, constructive ways.

Joint nesting is practiced by three species of anis and the related Guira Cuckoos. The most thoroughly studied of the anis are dull black, long-tailed birds with high-arched bills. That of the Groove-billed Ani has prominent longitudinal ridges and furrows on the upper mandible; that of the Smooth-billed Ani is more highly arched but lacks ridges. Together or alone, these two species are found over most of the tropics and sub-tropics of the western hemisphere. The larger species, the Smooth-billed, inhabits South America from northern Argentina northward, southern Central America, southern Florida, and the West Indies. The Groove-billed is less widely distributed in South America but occurs alone over the greater part of Central America and Mexico and in southern Texas.

Both anis are highly social, living in loose flocks that often contain ten to twenty-five birds, who vigorously defend their common territory against neighboring groups. Birds of wide ecological tolerance, they are equally at home in clearings in rain forest and amid cacti and thorny scrub, in pastures, plantations, gardens, and marshes, as well as dry savannas. They hunt insects, spiders, and small lizards among the foliage of trees, shrubs, and vine tangles, but chiefly in grassy or weedy fields, over which they run or hop with feet together, often leaping into the air to seize an escaping insect. They forage most profitably close to the heads of grazing cattle, horses, and mules, who stir up grasshoppers and other small creatures that the anis readily catch.

At the first nest of a Groove-billed Ani that I watched, belonging to a solitary pair, I was impressed by the way the incubating bird, put off the nest by my visit, flew, calling, directly to its mate and perched as close beside the other as it could press, while each billed the other's plumage. Larger groups rest in a long row, indiscriminately preening those within reach. At night they roost in an orange or other citrus tree, screened by dense, dark green foliage and protected by formidable thorns, or in a tangle of vines at a thicket's edge, all pressed together in a row and facing the same way.

As the breeding season approaches, Groove-billed Anis pair within the flock. When the rains begin, they start to build in a thorny tree, vine tangle, or arborescent cactus, at no great height. Several pairs may build a single nest, working in pairs, one member sitting in the growing structure and arranging sticks that the other breaks from trees or gathers from the ground and delivers, much in the manner of building pigeons. Two pairs may work simultaneously at the same nest. The completed structure is a rough, open cup of strongly interlaced twigs. It is lined with green leaves that are added daily as long as incubation continues and are not removed as they wither.

Even before building ceases, females begin to lay eggs with a soft, chalky white surface that is readily rubbed or scratched off, revealing a hard, blue or blue-green inner layer. Each produces about four eggs, which is the number often found in nests of solitary pairs. A nest of two pairs may contain eight; one of three pairs, twelve. Larger clutches are rare, as they cannot be efficiently incubated. One often finds intact eggs on the ground beneath an anis' nest, which remained inexplicable until Sandra Vehrencamp learned that, when ready to lay, the alpha, or dominant, female casts out all eggs already present. The lower-ranking females lay again but usually end with fewer incubated eggs than the alpha bird produces. This wasteful egg tossing is surprising in birds that appear so friendly, but it results in more nearly simultaneous hatching and probably increases the number of fledglings from a joint nest.

All the partners of both sexes in a joint nest take turns incubating, one at a time. No matter how few the minutes that an ani has been sitting it nearly always leaves when another arrives to replace it; the newcomer sometimes pushes into the nest beside a sitting bird that delays departure. The alpha male takes the dangerous nocturnal shift, when he is exposed to predation not only by snakes and terrestrial mammals but also by large carnivorous bats. After an incubation period of about thirteen days, the nestling anis, blind, black-skinned, devoid of down, escape from their shells. They are fed and brooded not only by parents who do not discriminate between their own progeny and those of collaborators, but, if they belong to a second brood, by juveniles of the earlier brood. A Smooth-billed Ani watched by D. E. Davis started to feed nestlings when forty-eight days old, and a Groove-billed watched by me at seventy-two days.

The parents, especially the males, defend their young with spirit. Voicing a menacing *grrr* and loudly clacking strong bills, Groove-billed Anis have repeatedly buffeted the back of my head while I inspected their nests. Juveniles a few months old join these defensive demonstrations. The nestlings are extremely precocious; when only six days old, with feathers just emerging from the tips of their long, horny sheaths, they jump from the nest when disturbed, to creep away and hide amid the herbage. After all is quiet, they climb up to their nest if convenient twigs help their ascent. At eleven days they are feathered and can make short flights from branch to branch.

Although in Sandra Vehrencamp's study, joint nests did not increase the production of young, they did promote adult survival, particularly that of subordinate females and of males who did not incur the high risk of nocturnal incubation. Alpha males, who alone were in charge of joint nests by night, suffered the highest mortality. However, by diminishing the number of nests in a population, joint nesting decreases the number of males incubating by night; and by dividing diurnal incubation among a number of collaborators, it increases the time that all group members enjoy for foraging and other aspects of self-maintenance. And it seems to satisfy the strong craving for companionship of these black birds that appear so affectionate.

We have looked at only a few of the many cooperative breeding associations but perhaps enough to call attention to their salient features. The first point to be emphasized is the amity that prevails among all group members. The rank order that assigns to each a definite status in the group, and is preserved by means so subtle that they are difficult to detect, minimizes internal conflicts. Food sharing, feeding of one grown bird by another, reciprocal preening, resting and sleeping in contact, displaying as a group, and in some species playing together are among the bonds that strengthen group solidarity. Internal amity does not contrast with external enmity, as among many human groups. Neighboring groups rarely fight but maintain territorial boundaries by displaying to each other across the often invisible dividing line—peace conferences that really keep the peace. The frequent migration of subordinate individuals from one group to another in which an opening occurs prevents debilitating inbreeding and helps to ally the groups.

The junior members of these groups live more safely than if they struck out on their own soon after they could support themselves, as commonly happens among birds that do not breed cooperatively. From their elders they learn to forage efficiently, to recognize and avoid enemies, and to build the more elaborate nests. By helping to rear broods that are usually younger siblings, they gain the experience that will make them more competent and successful parents when finally they disperse.

Cooperative breeding requires, and doubtless promotes, certain mental qualities. Group members not only recognize one another as individuals but remember relationships, as is necessary for incest avoidance, as noted in chapter 1. They must be able to cooperate and live amicably with associates. Plural breeders in groups with many simultaneously active nests, like Chestnut-bellied Starlings and San Blas Jays, appear to need a director to assign tasks to the numerous collaborators, so that no brood is neglected while others have an excess of attendants. No such direction has been detected; apparently, each co-worker sees where he or she can be helpful and serves accordingly, which seems to imply something like judgment. Cooperative breeding is so compatible with avian nature that one wonders why it is not more frequent among permanent residents. The reason appears to be that too much activity at nests of small birds, unable to protect their broods from enemies that they might attract, reduces their success. In regions where predators abound, birds tend to minimize their visits to their nests.

References: Brown 1987; Counsilman 1977; Davis 1940; Grimes 1976; Hardy et al. 1981; King 1980; Skutch 1959, 1983a, 1987b; Stacey and Koenig 1990; Vehrencamp 1977, 1978; Wilkinson 1982.

Chapter 4

✿

Emotions

DANCING CRANES, pairs of grebes displaying elaborately on a lake, albatrosses spreading great wings in nuptial demonstrations, male grackles addressing reluctant females with fluttering wings, fluffed plumage, and shrill cries all impress us as birds highly charged with emotion. Even naturalists reluctant to attribute much intelligence to birds recognize their emotional nature. Yet that a bird who appears to be highly excited is in fact emotionally stirred is the spontaneous inference of a sympathetic mind rather than a scientific conclusion. Intelligence is demonstrated by learning and profitable responses to unusual situations, whether in the wild or in a researcher's laboratory. The affections do not lend themselves to objective tests; we can only infer their presence from observed behavior.

The prolonged nuptial fidelity of many birds suggests that they are affectionately attached to their partners. Permanently resident Mute Swans, Jackdaws, Australian Ravens, Major Mitchell's Cockatoos, and Splendid Fairy-Wrens retain the same mates as long as both live; change of partners is rare in these species, as Ian Rowley reported. Konrad Lorenz believed that a male Graylag Goose who loses its mate never pairs again, as though unable to recover from his bereavement. Migratory or nomadic birds that separate after breeding have higher rates of "divorce"; but many rejoin their former partners in successive seasons, although whether they are drawn together by attachment to one another or to the territories they have shared is not clear. Similarly, with birds that remain together on the same territory thoughout the year, we are uncertain whether the mate or the territory is the stronger attraction.

It is widely held by professional ornithologists that male birds closely accompany their partners, especially while the latter are laying eggs, to prevent cuckoldry. Frugivorous birds who are not strongly territorial but wander widely as trees or shrubs offer abundant fruits now here, now there are more obviously held together by individual attachment alone. The many tanagers and other birds that travel in inseparable pairs through the months when their sexual urges are dormant appear to be held together by mutual affection. A long "engagement" period, often beginning when the birds are only a few months old and sometimes prolonged for more than a year before they breed helps to ripen a bond that will endure.

Birds do not choose their mates indiscriminately. If a male already holds a territory, the female's choice may be influenced by the quality of this real estate as well as by his particular attributes. However, birds often pair before they have territories, when only individual differences, often too subtle for us to detect, can determine their choices. Even in birds whose marital association is brief, as among ducks of the north temperature zone, the pair-bond can be strong while it lasts, as was demonstrated by Cynthia Bluhm's elaborate experiment with Canvasbacks reared from artificially incubated eggs. Nineteen pairs that had been freely formed served as controls. Ten other such pairs were separated and their members arbitrarily joined in different combinations. A third group consisted of twelve females who had been actively courted but had not indicated their choices of partners; each was given a drake who had not courted that particular duck. All the Canvasbacks were two, three, or four years old, and only one female had previously nested. Each dyad was placed in a cubicle with food, a nest box, and everything else needed for their health and comfort.

Of the nineteen strong, self-formed pairs, seventeen soon had eggs. Of the twelve females who had been courted but had not accepted mates, ten coexisted peacefully with the drakes assigned to them but they did not exchange courtship displays, synchronize their maintenance activities, or lay eggs. The females who had been separated from the males they had chosen became extremely aggressive toward their new companions, chasing and pecking them whenever they were in the water together. Although larger and heavier than the females, the long-suffering drakes did not retaliate. Five of them succumbed to this harsh treatment, although no male in the other cubicles died. Thus, although there was apparently no physiological impediment to breeding by all the dyads in the experiment, psychic factors inhibited mating and reproduction in all except the spontaneously formed pairs. Must not affection or some similar mental state be stronger in birds who pair for life than in ducks, who stay with their partners briefly?

Birds who live in pairs through the years have ways of expressing and strengthening their bonds. Wrens who forage amid dense vegetation where visibility is poor sing to each other, often antiphonally, with one member, usually the male, starting a verse that the other concludes. Their

Japanese Cranes

parts are often blended so skillfully that a listener not standing between them seems to hear only a single voice. Some continuously mated birds have greeting ceremonies. When the Black-striped Sparrows who hop over our lawn approach each other after a brief separation, they greet each other with a rapid flow of queer, whining notes, delivered simultaneously with a falling inflection, which I hear only on such occasions. As members of a pair of Vermilion-crowned Flycatchers come together, both flutter their wings while they deliver a short, high-pitched, somewhat trilled salutation, which they may repeat at intervals while they perch not far apart in a treetop. Mutual preening and gifts of food also bind mates together. These little courtesies seem less emotionally charged than the courtship of bigger birds who may pair only seasonally, but they may be expressions of enduring affection.

The contrasting stages of a nesting sequence contain many occasions for emotion. Some of the numerous New World flycatchers appear to be exceptionally emotional birds, or at least their various vocalizations seem more readily comprehensible to the human mind than do those of many other birds. While seeking sites for their nests, either sex sits in a promising spot in a tree or shrub and voices a prolonged flow of soft, twittering notes, the nest song. Among flycatchers from whom I have heard such twitters are the Vermilion-crowned, the related Gray-capped, the Sulphur-bellied, Golden-bellied, and Boat-billed. Sometimes the nest song stimulates neighboring pairs to rest in potential nest sites and sing. In the trees around our house, I have heard three species uttering these soft twitters at the same time. The emotions of these birds preparing to engage in what appears to be a satisfying activity are contagious, pleasantly exciting not only to their feathered neighbors but also to their human hearers.

Another bird who seemed to be emotionally stirred by the prospect of nesting was a Garden Thrush, who early on three consecutive mornings flew repeatedly to a certain spot amid the large red-and-green leaves of a *Taetsia* shrub in front of our dining room window. Here he stood for a minute or two while he sang a few subdued notes. After the third morning, he and his mate started a nest in the spot he had chosen. Although it is not usual for Garden Thrushes to help build the substantial, mud-walled bowl, he brought many billfuls of material to it. Similarly, a male Scarlet-rumped Tanager came again and again to sing in the bushy *Thunbergia* beside my study. After he had done this for several days, his mate went repeatedly to sit in the same spot amid the close-set twigs, while he perched, often singing, below her in the shrub. On the following day, she started a nest in the site he had chosen with song, and built it without his help, as is customary in Scarlet-rumped Tanagers. Although Buff-throated Saltators are usually silent, I heard one sing in a whisper while she collected material for her nest.

Incubation often impresses us as stolid, emotionless sitting that we would find excessively boring. However this may be with a broody hen who despite efforts to discourage her, persists in sitting for days on an empty nest, some birds give suggestions of emotional attachment to their eggs. Again, it is the flycatchers who most clearly reveal incubation to be more than the mere transfer of heat from the bare patch on the bird's abdomen to the developing embryo beneath her. While sitting for hours in a blind before a flycatcher's nest or just in passing by, I have often heard a subdued song that I have not heard at other times of the year.

After most of her absences to forage, a Vermilion-crowned Flycatcher voiced a little nest song as she passed through the doorway in the side of her bulky roofed nest and snuggled down on her two speckled eggs. Sometimes she twittered softly in the midst of a session of incubation and at intervals she called more loudly and was answered by her mate. While she uttered these notes, her usually hidden crown patch spread like a brilliant vermilion flower and seemed to illuminate her ample chamber. Flycatchers of many kinds keep their red, yellow, or white crown patches covered except when stirred by emotion. A less voluble Vermilion-crown entered her nest with a subdued nest song but rarely broke her silence while she sat. While she was absent, her partner, who like all male flycatchers never incubated, sometimes rested in the doorway with his head inside and rapidly repeated soft notes. He, too, appeared to be moved by the sight of the eggs.

Little gray Lesser Elaenias often appear to find incubation an irksome task. During their usually short sessions in their open cups, these flycatchers constantly turn their heads from side to side, looking around. Nevertheless, while sitting they may continue for minutes together a quaint little ditty. Other flycatchers that twitter softly while they incubate are the Gray-capped, Dusky-capped, and Boat-billed. Even the pesky Piratic Flycatcher, who calls so annoyingly while trying to usurp a covered nest built by some more industrious bird, utters a low, melodious, rhythmic murmur that may merge into a soft warble while she sits calmly on her eggs in a stolen structure. The vociferous Bright-rumped Attila, who has been transferred from the cotinga to the flycatcher family, hums a subdued nest song while she warms her three or four mottled eggs in a niche in a tree trunk or tucked amid massed epiphytes.

I have not heard a whispered nest song from a songbird (Oscine), but males and even females may sing more loudly while they incubate. For nearly two hours, a male Rufous-browed Peppershrike sat in a high, inaccessible nest without ceasing to repeat, at intervals of a few seconds, a loud, far-carrying song. Among the related vireos, males of the Yellow-throated and Warbling sing while taking their turns at incubation. The full-voiced songs of incubating male Rose-breasted and Black-headed grosbeaks guide ornithologists—and probably also predators—to their nests. Among female songbirds who sing loudly in their nests, in response to their mates or at other times, are Gray-breasted Wood-Wrens, Melodious Blackbirds, Yellow-tailed Orioles, and Blue-black Grosbeaks. In all these birds, the need to express feeling appears so strong as to overrule prudence.

When approaching their nests, some birds are needlessly active and noisy, moving around and calling loudly, expressing excitement or anxiety at the risk of attracting the enemies that they appear to fear. Even in predator-infested tropical forests, where prudent birds should return to their nests with silent directness, Blue-black Grosbeaks, Red-crowned Ant-Tanagers, and Black-hooded Antshrikes seem unable to repress their emotions as they revealingly return to their eggs or nestlings.

For some parent birds, hatching is an emotional event. One April afternoon, I set a stepladder beneath the small, mossy globe with a side entrance that a Mistletoe Flycatcher had built in a calabash tree in our garden. While I climbed up, the female clung with her head in the door-way, repeating notes so soft and low that I would not have heard them if she had not been so close above me. After continuing this soliloquy for possibly half a minute, she climbed inside and settled down with her head outward. She had been so absorbed by what was happening in the nest that she had not noticed me set the ladder below her; usually Mistletoe Flycatchers are more wary. Presently, noticing her visitor, she darted out and complained with the low, mournful notes usual on such occasions. Peering into the softly padded chamber, I saw a nestling in the act of hatching, with a little of its tiny pink body visible between the separating parts of the shell.

Sitting in her nest in a rose-apple tree at daybreak on an April morning, with a newly hatched nestling and a hatching egg beneath her, a Lesser Elaenia continued for many minutes to deliver a subdued, simplified version of her mate's dawn song. *De-weet, de-weet,* she repeated countless times, in a queer, dry voice. During the fortnight that she had incubated close beside an open window, I had not heard her voice such notes, which I believed were confined to males. The birth of a nestling stirred her to utter notes that she seldom used. While a Boat-billed Flycatcher incubated in a guava tree, also close to the house, I did not hear the nest song that had been fre-quent while she chose the site and built her bulky, open cup; but the day a nestling hatched she was moved to utter these soft notes again.

It might be surmised that these vocalizations just as a nestling is born serve to advise the father of the event, so that he will promptly begin to feed it; but careful studies yielded no evidence for this. These cozy chants appear to be simply expressions of strong feeling. Male birds who do not incubate may also respond emotionally to their first sight of their nest-lings. Males of both Vermilion-crowned and Gray-capped flycatchers used soft notes not unlike the nest song when first offering food to hatchlings, whom the males did not discover until a few hours after their mates had begun to feed the young.

The next strongly emotion-stirring event in an undisturbed nesting sequence is the departure of nestlings that for weeks the parents have labored strenuously to nourish. Returning with a laden bill, a Vermilion-crowned Flycatcher—the mother, I believe—found that a nestling who had lingered unusually long in its domed nest had flown out in her ab-sence. For many minutes she continued a flow of soft notes much like the nest song, at first while alone, then after she had joined her fledgling in a neighboring tree. Similarly, the departure of a fledgling Gray-capped Fly-catcher prompted many repetitions of the parents' nest song, which I had not heard during the long days preceding the youngster's exit from the nest.

Long-billed jacamars are, for nonpasserines, exceptionally songful birds. One of the most elaborate celebrations of a fledgling's graduation from the nest that I have witnessed was by a pair of Pale-headed Jacamars in Venezuela. Emerging from its long natal burrow in a high bank, the young jacamar rose to the crown of a roadside tree, where both parents promptly joined it and started to sing. Each crescendo of *weet*'s and twitters led up to a high, thin trill. With variations, they repeated this song again and again, while their bodies turned from side to side and their tails twitched rapidly up and down, beating time to their animated notes. At intervals the fledgling joined in with its weaker voice, flagging its tail as the parents did. For a long while, the parents continued to sing what impressed me as a triumphant paean for the successful conclusion of their nesting. The exit of a second fledgling later that same morning was greeted with song by the single parent then present, but the celebration was briefer, as though the earlier performance had drained the jacamar of emotion.

The emergence of two young Rufous-tailed Jacamars from a burrow in a Costa Rican bank was also the occasion for much singing by their parents as well as the fledglings themselves, who while still in the nest had practiced weaker versions of the adults' elaborate songs. Southern House-Wrens, unquenchable songsters, sing more profusely while their fledglings timidly sever contact with the gourd, nest box, or cranny in a tree or building where they grew up. Birds as different as Black-crowned Tityras, who raised three nestlings in a cavity high in a tree, and Marbled Wood-Quails, who hatched four chicks in a covered nest on the ground, seemed unusually excited when their young flew or walked from their respective nests.

Nests where a brood was raised may continue to excite strong parental feeling. During the week after the departure of their two fledglings, both parent Gray-capped Flycatchers went repeatedly to their deserted nest, singly or together, to deliver nest songs or sundry spirited or bizarre sequences of notes from their varied vocabulary. Occasionally, they brought food to the empty nest before taking it to their progeny outside. Even a full month after the young birds flew, I watched both parents sing loudly near the nest. During all this interval, they dutifully attended their young in neighboring trees. This continued interest in the old nest was not in anticipation of using it for a second brood, which Gray-capped Flycatchers rarely attempt in this region. Likewise, male Scarlet-rumped Tanagers sometimes sing beside nests where their young were reared. For over a month after the fledglings left their nest amid the variegated foliage of a croton shrub beside my study, a male, doubtless their father, came almost daily, on some days twice, to sing profusely while standing in the empty bowl, on its rim, or close beside it. Forty-six days after her first brood flew, his mate started a new nest a few yards from the old, disintegrating structure.

Pale-headed Jacamars (immature, at left, and two adults)

When their nests are approached by an animal that might harm them, birds behave in contrasting ways. Some flee at the first sign of danger, trying to avoid detection, a tactic that in some circumstances may best promote the survival of both parent and progeny. Others remain until the potential enemy has come near, then try to lure it away by simulating injury (see chapter 9). Still others, prompted by parental affection that overrules prudence, boldly attack with pecks, nips, or blows of the wing any animals too powerful to deter. On a number of occasions, I might, if so inclined, have caught or killed the small bird who confronted me. Once I rescued a female Scarlet-rumped Tanager hanging by one wing from the jaws of a snake she had vainly tried to drive from her eggs. Such supererogatory zeal for progeny could hardly be promoted by natural selection, for the loss of the parent would in most cases be followed by loss of the brood. Likewise, the futile cries that many birds utter while fluttering around a nest that is being pillaged by a predator or inspected by a naturalist are expressions of parental distress without survival value.

It is remarkable how often the sounds that birds make suggest the emotions that we might feel in similar circumstances: soft notes like lullabies while calmly warming their eggs or nestlings; mournful cries while helplessly watching an intruder at their nests; harsh or grating sounds while threatening or attacking an enemy; sharp, castanetlike clacking of the bill while trying to intimidate a rival or interloper. Birds so frequently respond to events in tones such as we might use that we suspect their emotions are similar to our own. Some birds, however, appear to lack notes appropriate for harassing occasions; when their nests are threatened they can only sing or emit melodious sounds, as did Black-cowled Orioles, Melodious Black-birds, and White-collared Seedeaters when I inspected their nests.

The indispensable activities of breeding birds—building nests, incubating eggs, feeding and brooding young—fellow innate patterns and give no clue as to accompanying feelings. While watching them we might infer that the birds mechanically follow a course programmed in their genes, were it not for embellishments such as crooning to eggs or young, exchanging songs with mates, or anticipating nestlings. The unessential illuminates the essential. Watching nests of many diverse birds over six decades has left me with the strong conviction that many of them, perhaps all, are affectionately attached to their families. Irrefragable proof of this I cannot give because the inner lives of all creatures, including those most like me, are hidden behind an opaque veil, and I have only inference and intuition to enlighten me.

References: Bluhm 1985; Rowley 1983; Skutch 1960, 1976.

Chapter 5

❦

Play

PLAY, PROPERLY so-called, is spontaneous, intrinsically rewarding activity, a pastime in which healthy animals who have satisfied all vital needs expend excess energy for enjoyment alone, with no ulterior motive. Whatever extraneous benefits play may yield—as in strengthening muscles, sharpening skills, or improving social relations—are incidental, not its ends. Much of animals' play is the exercise of their special endowments for pleasure rather than for survival or reproduction, as when well-rested horses gallop over their pasture, dolphins race in front of an advancing ship, birds soar on widespread wings, or birds sing simply to amuse themselves. These same endowments are also indispensable for the life-sustaining occupations for which they evolved, and the close similarity of obligatory and optional activities often makes it difficult to distinguish play from other behaviors. The occurrence of play among the orders of birds and its types—locomotor, object, or social—were surveyed by J. C. Ortega and M. Bekoff.

When young birds still fed by their parents pick up, toy with, then drop such inedible objects as bits of leaf and bark, are they playing or trying inexpertly to satisfy their hunger? Similarly, when they fiddle with nest materials, are they amusing themselves or clumsily expressing the first stirring of the nest-building instinct? Much apparent play may be "vacuum" activity, the release of innate drives in inappropriate situations, as when birds indulge in courtship displays in the absence of the other sex. This makes the borderline between play and nonplay difficult to draw. The more an activity differs from the essential business of escaping perils, finding food, and reproducing, the safer we are in calling it play, the greater the probability that the animal engages in it simply for enjoyment. In the following pages we shall give special attention to such activities, trying to avoid those which merely simulate play.

For insight into the psychic life of animals, the study of play is of first importance. Unlike the indispensable life-sustaining activities, it appears to escape from the rigor of natural selection into the realm of spontaneity and freedom. Instead of being innate modes of behavior widespread in a species, some kinds of play are rare or have been so seldom reported that they appear to be inventions of active minds, perhaps imitated by companions. It is significant that play is largely, if not wholly, restricted to the two classes of vertebrates with the most developed brains—mammals and birds—although it has been attributed to certain fishes and fiddler crabs. The play of mammals has received more attention than that of birds, and many theories, both physiological and teleological, have been offered to explain it. They may account for activities that resemble play, but, with Fraser Darling, I believe that animals play "from an innate spirit of fun."

Young altricial birds, nourished by their parents, may be more inclined to play than are precocial chicks, who must devote their energies to finding food for themselves. House Martins, who like other swallows remain in the nest longer than do most small birds, "have fun even in the cradle." Len Howard "watched a young one stretch half its body out of the nest to play with a smaller Martin in a neighboring nest. Twittering excitedly with head craned forward, it just managed to touch the tip of the other baby's beak. The smaller one entered into the fun and bobbed up from its nest to take the kiss, both withdrawing directly afterwards. . . . The game continued for some time."

After they leave the nest and are stronger, young birds are more playful. When only seventeen days old and about a week out of the nest, Margaret Nice's hand-reared Song Sparrows began to frolic. They ran over the floor with sharp turns, flapped their wings, jumped from spot to spot, and playfully rushed at their siblings. In the following days, their frolics became livelier. An American Redstart likewise started to gambol at the age of seventeen days, flapping his wings and flying madly around his cage. From the older literature, Nice gave examples of similar behavior of doves and newly fledged Sedge Warblers. Her hand-raised Northern Bobwhite would flap his wings, crouch, and suddenly make little flights.

Howard described the gambols of two fledgling Great Tits, who touched bills in a quick, flying chase. In another game, one of these fledglings perched high on an ascending twig while the other, with much twittering, slowly sidled up to him from below, swinging his body from side to side as he advanced. When he reached his playmate, he touched the latter's wingtip and they flew around in a whirling chase, then alighted to repeat the game, taking turns to be top bird. She gave a vivid account of a young Eurasian Blackbird in her garden amusing himself by tossing an old walnut shell a foot or more away, then pursuing it with wing-flicking

Eurasian Blackbird (immature)

leaps, or throwing it over a low bush and flying around to retrieve his plaything and continue his game.

Recently I watched Tawny-winged Woodcreepers at play. A great fallen frond of a chonta palm was draped over a branch in the undergrowth of the rain forest, its tubular sheathing base hanging upright. Two, three, and occasionally four of the brown birds flew together to the open upper end of the tube, entered it, and sometimes one climbed tail first down into the cylinder, out of sight. Soon they flew away, to return a few minutes later. For at least half an hour, they went again and again to the tube, attracted perhaps by its resemblance to cavities in which these woodcreepers nest and sleep. Although well grown, they appeared to be juveniles; adult Tawny-winged Woodcreepers are such unsocial creatures, the females nesting quite alone, that I doubt that they would bunch peacefully together at the entrance to the tube, as these birds did. They were youngsters enjoying a frolic.

Among nonpasserines, three Fischer's Turacos raised by R. E. Moreau in Africa were "exceedingly playful in a monkey-like fashion." When about one month old and still incapable of flying, one of them would jump from the ground to seize the leg or tail of another on a low perch and try to jerk him off. If he succeeded, the second bird would play the same trick with the first. Then they would dash up and down their perches, chasing one another wildly. They were also very playful with their human friends. They loved to be touched, stroked, or tickled and in return made mock attacks that were often quite alarming. Their playful disposition persisted into adult life. After they could fly, one turaco would pull off Moreau's spectacles with a rapid twitch while perching on his shoulder, then fly off with a chortle of satisfaction. Two Gray-headed Chachalacas we raised with the help of a broody hen from eggs abandoned by their mother were also playful. One would chase the other two or three times around a bush, then they would run over the lawn to repeat the performance around another shrub. After their games, they would rest close together and they regularly roosted in contact.

Only a strained interpretation would view these juvenile amusements as forerunners of adult aggression or the precocious manifestations of innate, life-preserving patterns of behavior, such as predator avoidance. On the contrary, they appear to be spontaneous expressions of youthful vitality and exuberance, like many human children's games. Such behavior is most frequently witnessed in hand-reared birds or those so closely associated with human benefactors that they have lost all fear of people; reports of play by fledglings in the wild are hard to find. This may be because young birds well fed by people have more excess energy to discharge in frolics than do dependent young in natural surroundings, and also because vigilant parents tend to keep their fledglings hidden from enemies, which too frequently include people. It would be surprising if precocious playfulness were restricted to individual birds closely associated with humans.

A simple form of amusement of older birds is dropping things and watching them fall. While a schoolboy, Edmund Jaeger watched six or eight House Sparrows carrying pebbles to the edge of a long, sloping, gravel-covered roof of a two-story building and releasing them. As the pebbles fell, each sparrow turned its head sideways to watch with one eye and perhaps better to hear the *ping* as the stone struck a sloping wooden door that led into the basement. In later years, Jaeger was inclined to doubt the accuracy of his recollection until, long afterward and in another state, he watched two House Sparrows drop small bits of crushed stone from the roof of a building onto a cement pavement.

Len Howard's Great Tits uncovered a glass jar containing seashells, picked them out, and with sharp twists of their heads tossed the shells either down on the floor or across the room, intently watching their fall. These tits treated matches in similar fashion, strewing them all over the room, along with the matchbox they had pulled open. An unfamiliar Jackdaw entered her bedroom by descending the chimney, alighted on her dressing table, and skidded about on the polished surface while he dropped small articles to the floor, with sideward-turned head and an intense expression watching them fall.

In Australia, as A. J. Marshall reported, a male Spotted Bowerbird, rearranging the ornaments on his display platform, picked up shells from the heap and dropped each twice or thrice in the same spot, apparently because he enjoyed hearing them tinkle against one another. Gulls and Ravens drop things for a more serious purpose when they break the shells of mollusks or sea urchins by casting them from a height onto rocks or other hard surfaces, and Egyptian Vultures crack open Ostrich eggs by throwing stones upon them. It is not likely that playful dropping by sparrows and tits is derived from the practices of these larger birds, but could these habits have developed from the amusement of watching objects fall?

Some birds make the game more active by catching things they drop. On an August morning, a Boat-billed Flycatcher flew into a tree in front of our house in Costa Rica with a big brown feather in its bill. After beating the feather against a perch, as though it were a large insect, the bird dropped the plume. As it floated slowly downward, the flycatcher's mate darted out and caught the feather, carried it to a tree, and knocked it against a branch. Soon the second bird also dropped the feather, only to dart out, catch it as it wafted away, then strike it against the branch again. This bird carried the plume to another tree and continued to beat it but not as vigorously as might have been done with a cicada, a favorite food. Finally, the flycatcher released the feather, and the pair flew away together.

Len Howard watched Barn Swallows at play over a grassy Devonshire hillside where ducks and geese had shed white feathers. A swallow would drop down, seize a feather in its bill, then swoop upward to circle above the other swallows and drop the plume. As it floated down, another caught it. So the game continued while with marvelous grace the birds traced wide arcs through the air. In South Carolina, Alexander Sprunt, Jr., watched a similar performance by a male Purple Martin. This time the actor was alone and the plaything was a wisp of grass. From high in the air, the martin would drop the straw, then, diving, sideslipping, and rolling, would plunge beneath it and seize it in his bill as he zoomed upward to meet it. For nearly a quarter of an hour, the bird continued this play, performing wonderful aerial maneuvers, including nose dives, falling leaves, and stalling at the top of an upward climb, then gliding backward, tail first, for a considerable distance. The martin even turned completely over to sail along upside down for several yards. The playful martin performed most of the stunts known to airplane pilots.

Common Ravens carry aloft twigs or sprigs of heather, to catch them as they fall. On an island off the coast of Peru, Philip Ashmole and Humberto Tovar watched about twenty Inca Terns who had recently begun to fly snatch fragments of seaweed or other small, inedible things from the surface of the ocean, fly a short distance with a piece, drop it, then frequently catch it as it fell, sometimes as much as ten times with

Barn Swallow

the same object. Several young terns might indulge simultaneously in this play, either competing for the same piece or plucking another from the water. On Ascension Island, Bernard and Sally Stonehouse watched frigatebirds play by snatching feathers or strands of seaweed from one another in midair. In addition to amusement, these aerial games of terns and frigatebirds may improve skills used by foraging adults.

Other birds disport in the air without playthings. One evening, strolling along a shingle beach on the coast of Maine, I watched Herring Gulls enjoy the strong onshore breeze. Rising high, they headed into the wind to hover motionless on outstretched wings. All along the shore was a line of gulls, rising, descending, or balancing themselves in the breeze. They delighted in gliding swiftly downward with the wind, then using their great momentum for an upward swoop. The pleasure of riding effortlessly high in the air appeared to be the only motive for this relaxing play, before in the waning light they struggled against the headwind toward the islet where they nested and slept.

Konrad Lorenz told of the play of his Jackdaws at Altenberg in an autumn gale. Dropping heavily as rocks, permitting the breeze to toss them high as they "fell" upward, then with a casual flap of a wing turning themselves over and diving faster than a falling stone, they played effortlessly with the wind, masters of the gale that appeared to be tossing them about. These maneuvers were not purely instinctive actions but carefully learned movements perfected by practice and repeated for enjoyment alone.

The aerobatic play of Common Ravens is well known. While flying, they rotate sideward about their long axis, sometimes through a half-roll of 180 degrees, sometimes through a full roll of 360 degrees. For a half-roll, the bird partly folds its wings, rotates rapidly onto its back for about one second, then reverses its movement to regain the upright posture while spreading its wings. It may revolve either to the right or to the left, most often the latter. Dirk Van Vuren, who made a special study of this behavior, watched a raven roll nineteen times, in both directions, in one continuous performance. Far more rarely, a raven made a full roll, with one steady motion until it regained its normal posture. Occasionally, a bird made two continuous revolutions about its long axis.

These ravens rolled while flying alone or in parties of two to five, and as frequently in fall and winter as in spring. On rugged, windy Santa Cruz Island off the coast of southern California, ravens rolled much more frequently than on the mainland. If rolling entered importantly into courtship, one would expect it to be most frequent in spring, and no less common inland than on the island. Evidently, ravens roll because they enjoy the exercise. Quite different is the ravens' vertical spin when, diving from a great height, they revolve more and more rapidly around their longitudinal axis, repeating this maneuver over and over.

Even sombre vultures, who appear so staid and solemn, seize an opportunity for playful flight. While a violent gale from the north blew over Cerro Verde in El Salvador, Walter Thurber watched a dozen or more Black Vultures soaring in the updraft above the windward slope to as much as 1,600 feet (about 500 m) above the mountain's summit. At intervals these birds dived precipitously, with the wind behind them, as much as 2,300 feet (700 m) down the mountain slope. A group of three made about twenty-five dives, in V-formation or in a line, with impressive speed and shrill sounds of air passing through their wing plumes. At the end of a dive, they used their forward momentum to circle around the lee slope into the updraft again, all, it appeared, just for the exhilaration of swift movement.

As swallows gather from afar to enter a communal roost, they join in compact flocks of hundreds or thousands of individuals, which swing here and there high above the treetops, the mass of birds constantly changing shape like a huge aerial amoeba, with internal particles in constant agitation. The mass drops toward the roost site in an umbrageous tree or field of sugarcane, only to rise again and continue to drift high in the air. Finally, as daylight grows dim, a cone-shaped protuberance extends from the underside of the mass as the swallows shoot downward in a contracting stream, as though poured through an invisible funnel, to vanish amid the foliage where they will sleep. Southern Rough-winged Swallows and other tropical species go to roost in this spectacular fashion, and Chimney Swifts approach and enter a tall stack in much the same way. These vespertine maneuvers certainly do not make communal roosts harder for predators to find. The birds seem to enjoy an aerial frolic before they retire.

The social gambols of Black Guillemots on the coast of Ireland were vividly described by Edward Armstrong, who called these white-winged relatives of auks and murres "playboys of the western world." From all over a bay they rise with one accord and fly around with coordinated movements, their bodies fringed with a quivering white halo on either side. They swim, splash, and dive together and, when tired of their play, the whirring host alights to stand or squat on a pier, like penguins in a long row. Armstrong believed that these frolics are derived from courtship maneuvers that the guillemots continue throughout the summer because they are so enjoyable.

Acrobatics as well as aerobatics amuse Common Ravens. Among their less frequently reported antics is hanging beneath a perch. In Nova Scotia, Richard Elliot noticed a raven dangling by one foot beneath an exposed branch. The black bird grasped the branch with its bill and released the foot, to support all its weight by the bill alone, without the aid of its partly open, motionless wings. Next, the acrobat removed its bill from the branch to remain suspended by both feet, the mirror image of a normally perching raven. For several minutes it continued these stunts, alternately hanging by one foot, two feet, or only the bill. One of the ravens looking on tried to copy the trick but, unable to sustain its weight by its bill alone, flapped its wings to prevent falling. Whereupon the first raven, croaking loudly, supplanted it from the branch and performed as before, as though to demonstrate the superior strength of its mandibles.

The colorful Galah, one of Australia's many parrots, has a similar habit, hanging by its feet upside down from branches, telephone wires, or television aerials, and swinging to and fro with spread wings, while screeching as though it enjoyed this stunt. Both adults and young behave this way, especially between feeds in fine weather, when a number of flock members may join the first performer. Except for the Galahs' relaxed, unfluffed plumage and the weather, these acrobatics are very similar to those of the rain dance, which is started by the first drops of a shower. Ian Rowley, who

described this play, also told how Galahs slide down the guy wires of television towers. These social birds seem to enjoy their lives much more than
most animals do. "They appear to get real pleasure from the perfection of
their flying, swerving in and out of trees with a consummate skill quite
superfluous to the mundane need of commuting."

We need only watch schoolchildren at recess on their playground or
boys and girls sledding on a snowy slope to be convinced that sliding
swiftly down an inclined plane is great fun. Nonhuman creatures enjoy
this sport on the rare occasions when opportunity arises. In A. C. Bent's
life histories of North American birds is a detailed account of Common
Ravens coasting down the high, crumbling bank of a river. A dozen at a
time would stand, either sideways or facing upward, while they slid down
amid rolling pebbles and clay, expressing their enjoyment with harsh croaks
audible afar. While some ravens slid, many others watched from surrounding trees, applauding the sport with their cries, and taking their
turns as the others tired of it. On a much smaller scale, Great Tits enjoyed
slipping down the slopes of Len Howard's pillow, while she lay ill in her
bed, "a miniature alpine sport."

An aquatic version of this game was played in Iceland by Common Ei-
ders, whom B. B. Roberts watched shooting rapids, then repeatedly walking up along the banks to enjoy again the exhilaration of another ride on
the swirling water. In California, a female Anna's Hummingbird com-
bined aerial with aquatic play, as seems proper for so airy a creature. While
Emerson Stoner watered his garden with a solid stream from a hose, she
attempted to alight on it as though it were a twig or a branch and, sitting
transversely, rode down the flow, repeating this stunt over and over. Years
later, Ernest Callenbach watched a similar performance by a hummingbird, as he told me in a letter.

Birds, like children, enjoy a ride on almost anything. G. Murray Levick
vividly described how Adélie Penguins amuse themselves in Antarctica:

They would spend hours playing a sort of "touch last" on the sea ice near
the water's edge. They never played on the ground of the rookery itself, but
only on the sea ice and the ice-foot and in the water, and I may here men-
tion another favourite pastime of theirs. I have said that the tide flowed past
the rookery at the rate of some five or six knots. Small ice-floes are continually drifting past in the water, and as one of these arrived at the top of the
ice-foot, it would be boarded by a crowd of penguins, sometimes until it
could hold no more. This "excursion boat," as we used to call it, would float
its many occupants down the whole length of the ice-foot, and if it passed
close to the edge, those that rode on the floes would shout at the knots of

penguins gathered along the ice-foot who would shout at them in reply, so that a gay bantering seemed to accompany their passage past the rookery.

Arrived at the farther end, some half a mile lower down, those on the "excursion boat" had perforce to leave it, all plunging into the tide and swimming against this until they came to the top again, then boarded a fresh floe for another ride down. All day these floes, often crowded to their utmost capacity, would flow past the rookery. Often a knot of hesitating penguins on the ice-foot, on being hailed by a babel of voices from a floe, would suddenly make the plunge and all swim off to join their friends for the rest of the journey, and I have seen a floe so crowded that as a fresh party boarded it on one side, many were pushed off the other side into the water by the crush.

Who, reading this, can doubt that the penguins thoroughly enjoyed their rides?

"I'm king of the castle" is a game frequent among young deer and other ungulates, who successively try to supplant one another from a hillock or other slight elevation. In the corresponding sport among birds, a branch or other perch takes the place of the hillock. Howard's Great Tits and Moreau's turacos provide examples of this game as played by fledglings. Adult birds also indulge in it. For three years, at all seasons, two Eurasian Blackbirds in Howard's garden engaged in it almost daily. If one failed to appear, his playmate would go in search of him and lead him back to the tree stump that served as their "castle." For hours together, with slight variations, the two alternately stalked each other and stood upon the stump in a cocky attitude. This was not a territorial dispute, as the stump was on unclaimed ground.

Some of the most elaborate social play is frequent among cooperatively breeding birds who live in closely knit families of parents with young of different ages. One of the simpler of these games is played, mostly in the early morning and late afternoon, by Brown Jays, including full-grown, nonbreeding individuals. One jay, standing in a tree in front of another member of the family, stretches up its legs and makes feints with its bill at the other, now here, now there, bobbing up and down, twisting and turning from side to side in a spirited manner, until it appears a feathered clown. The bird so assailed with harmless thrusts turns its beak toward the other and erects the feathers of its head and upstretched neck, looking very bizarre as it silently endures the feigned attack. Later, the roles of the two participants may be reversed, the attacker now assuming the passive attitude. The jays were already nesting when I watched these antics, which appeared to be play rather than courtship.

Toucans of tropical America play with their great, colorful beaks. Fiery-billed Aracaris, middle-sized toucans who live in family groups, strike their hollow bills resoundingly against a trunk or branch, apparently only to hear the report. In a high treetop in or near the forest they engage in a game. One that I watched began when two aracaris, facing each other, struck their long beaks together. Then they grasped each other's bills and pushed until one was forced backward and hung briefly below the limb, after which it admitted defeat and withdrew a short way. A third member of the flock now advanced to challenge the victor on his perch. Again the opponents knocked their bills together, grasped, and pushed. This time the winner in the first bout was forced from the bough and retired, leaving the newcomer as uncontested champion. These contests appeared to be wholly friendly; the loser was never threatened or pursued.

Southern Ground Hornbills, largest members of this Old World family, stalk in groups over the savannas of Africa south of the Sahara, hunting small vertebrate and invertebrate animals. In the Kruger National Park in the Republic of South Africa, A. C. and M. I. Kemp watched immature and adult birds play, often by grasping each other's long, low-casqued bills and wrestling, much in the manner of toucans, to whom hornbills are not closely related. They tossed things about with their beaks, chased one another on foot or in the air, jumped on their companions' backs from a bank, tugged at one another's legs, or mischievously pursued other birds such as Helmeted Guineafowl.

In India, Anthony Gaston watched Jungle Babblers engage in two forms of vigorous play. The first, which he called "rough and tumble" was a tussle between two or more birds, some of whom lay more or less passively on the ground while others rolled on top of them and pecked them deliberately but gently. Most of the four or more participants in these scuffles were under a year of age; breeding adults were never seen to join in. In "mad flights," the second kind of play, one bird or several flew rapidly and apparently aimlessly, twisting and turning wildly through the branches of a tree. Other babblers, including the Gray-crowned, Large Gray, and Arabian, all cooperative breeders, play similar vigorous games.

When they leave their roost in a small desert tree, Arabian Babblers also perform what Amos Zahavi called their "morning dance" on the ground beside a bush. The dancers, who may include the whole group of adult and immature birds, alternately stand in a row and gather into a tight ball, while individuals try repeatedly to force themselves into the middle of the row or the center of the clump. Sometimes they continue this play for half an hour. When they find water, these babblers drink and bathe, then indulge in a "water dance" not unlike their morning frolic. This is hardly a performance that helps dry their plumage, for by bunching together they delay evaporation. Moreover, even individuals who have not bathed join in the ritual after bathers have started it.

Among the more playful of feathered creatures are Australia's gray-brown, cooperatively breeding Apostle Birds, so-called because they live in groups that sometimes contain twelve, the number of apostles in the New Testament. When cooled by a drop in temperature in summer or on a warmer day in winter, these mudnest builders indulge in gambols that have been described by Merle Baldwin. They hop rapidly up a ladder in a garden only to glide down, repeating this performance again and again. They follow a leader around a tree, each crouching low and trying to nip the tail of the bird ahead. They dart away, to return calling loudly. In a rougher game, they strike out with their feet and try to turn a companion on its back. Or the dominant male may voluntarily lie on his back in a relaxed attitude while other birds peck his abdomen, until he jumps up and leads another chase around a tree or shrub. Apostle Birds pull twigs to the ground and hold them with their feet, while with their bills they tug and twist until the branches break. They strip off leaves and tear apart flowers. A young male may challenge the leader of the group by approaching with a leaf held high in his bill. The other seizes the leaf, and the two pull in opposite directions until it is torn apart. Successive tugs-of-war reduce the leaf to ever smaller fragments.

Birds, like children, sometimes engage in rough or destructive play. Tame American Crows pull apart papers, household articles, and colorful flowers in the garden. Howard's Great Tits tore open the toy animals she gave them, pulling out the stuffing. This propensity to pull things apart may yield nourishment, as when English tits discovered that they could tear off the caps of milk bottles left on household doorsteps and drink the cream, a habit that spread rapidly through the population, probably by imitation of the enterprising discoverer(s) of this bounty. Also like human children (and too frequently adults) sportive birds are not always considerate of other creatures. Young Willow Warblers and Chiffchaffs delight in chasing other swiftly flying birds, often much to their annoyance. These pursuits do no great harm; sometimes the pursued joins in the spirit of the game and when it tires and perches, the active warbler seeks another bird who will consent to be chased. But when ravens charge into a compact flock of small birds, terrifying and scattering them, perhaps injuring or killing some, the "sport" turns ugly.

Vocal organs, no less than limbs, may be exercised for amusement. Singing for enjoyment alone, whether by birds or humans, is a form of play. The first songs of immature birds, their "subsongs," which may differ greatly from the songs of adults of their kind, fall into this category. In a Costa Rican thicket, I listened enchanted to the sweet, diffuse, rambling monologue of a Plain Wren, still fed by a parent. I thought it a pity that when the bird grew up this charming recital would be replaced by the

stereotyped *chinchirigüí* that paired Plain Wrens sing antiphonally, the female adding *güí* to her partner's *chinchiri*. The reason for this change is
that adult songs serve to reveal specific identity, as juvenile ramblings,
which may be much the same in different species, do not.

The songs by which adult birds advertise their possession of territory
and attract mates serve an important biological function and cannot properly be called play, although, when elaborate, they may include a modicum
of playfulness. Mimetic song, or chatter, often appears to be play. Whether
Marsh Warblers' impressive repertoires of borrowed themes (chapter 2)
fall into this category I am uncertain. They are sung to attract mates but
they appear to be superfluously elaborate and they are repeated by warblers wintering in Africa, when reproductive activities are in abeyance.

In the top of a tall pine tree in the Guatemalan mountains, high above
the tangle of blackberry bushes where his mate incubated two lovely blue
eggs, a Blue-and-White Mockingbird sang delightfully, with notes ranging from deep, mellow whistles to light, airy trills. This was his true song,
undiluted by imitations while he sang early in the morning. As the hours
passed, he appeared to tire of classical music and turned to nonsense songs
with many harsh and churring notes. In the dim light when he retired to
roost in a thicket, again when he awoke at dawn, and at intervals through
the day, the blue mockingbird chattered in a most amazing fashion, min-
gling shrill squeals, guttural croaks, whistles, warbles, trills, and imitations
of other birds in the most fantastic vocal hodgepodge that I have heard.
These prolonged recitals are heard months away from the breeding season
and appear to have nothing to do with the serious business of reproduction. I could conclude only that the mockingbird was playing with his supremely flexible voice, finding pleasure in demonstrating its range and
power to himself, if not to others.

Although woodpeckers have well-developed voices, they tap or drum
in situations where other birds would call or sing. When they find a metal
gutter or wooden siding on a house, they can hardly resist beating tattoos
so loud and long that they distress householders who sleep late. When a
female Red-crowned Woodpecker was about to enter the tree hole where
she slept, she discovered that the hard white diaphragm across the end of
a broken branch of a cecropia tree responded resoundingly to her tapping.
As though enchanted by her discovery, she continued for several minutes
to drum on it, while her mate looked out of his bedroom above her. She
appeared to tap just to hear the noise, as children often do.

Probably birds, like many of us, derive a greater sum of happiness or
contentment from the efficient performance of life's necessary tasks than
from occasional diversions. Foraging, building nests, and attending young
may be pleasant occupations so long as the birds are not overworked. But

the evolutionary biologist and the behaviorist may contend that birds' indispensable activities are programmed in their genes and that it is no more necessary to postulate enjoyment, or even awareness of what they are doing, than it is to suppose that a machine enjoys or is aware of performing the operation for which it was designed. However, a bird must use its mind to adjust its instinctive activities to the endlessly variable circumstances of the environment, as by selecting a territory and nest site from a number that might be available to it, decisions no machine is ever expected to make.

When we turn from animals' basic, life-preserving activities to contemplate their play, we find a very different situation. Like everything that we or any other creature does, play must have an innate foundation, which appears to be the satisfaction that the unconstrained exercise of muscles and limbs brings, the pleasure that many animals appear to find in swiftly moving or being moved, as in running, swimming, flying, riding, or being borne by the wind. Given this background, the details of play are so variable, so dependent upon external circumstances, that they could hardly be genetically determined.

It is highly improbable that natural selection could promote such activities as sliding down a smooth inclined plane, repeatedly shooting rapids, or riding an ice floe or the jet from a garden hose, when such sport contributes nothing to birds' survival or reproduction and requires situations rare in their lives. The play of birds is diverse, unpredictable, and opportunistic. Often it appears to be started by an inventive individual who detects an occasion for enjoyment and is joined by companions. Only conscious beings are capable of enjoyment. For settlement of the moot question of animal awareness, play is decisive. Of all our reasons for believing that birds are feeling creatures who find satisfactions in their lives, their frolics are the most convincing. It is pleasing to know that birds, who give so much pleasure to us, are themselves capable of enjoyment.

References: Armstrong 1940; Ashmole and Tovar 1968; Baldwin 1974; Bent 1946; Callenbach, in letter; Darling 1937; Elliot 1977; Gaston 1977; Howard 1952, 1956; Jaeger 1951; Kemp and Kemp 1980; Levick 1914; Lorenz 1952; Marshall 1954; Moreau 1938; Nice 1943; Ortega and Bekoff 1987; Roberts 1934; Rowley 1990; Skutch 1983a, 1983b, 1987a, 1987b, 1989; Sprunt 1944; Stonehouse and Stonehouse 1963; Stoner 1947; Thurber 1981; Vuren 1984; Zahavi 1990.

Chapter 6

❧

Counting
and Timing

FOR AN INTIMATE study of nesting birds, it is often necessary to conceal oneself in a blind (which the British call a hide), a little tent or wigwam of cloth or a more solid construction, with narrow windows through which the watcher can peep. Believing that birds could not count, even to two, some ornithologists or photographers would approach their blind with a companion who, after the watcher was ensconced within, would walk conspicuously away, thereby, it was believed, making any bird who happened to be looking behave as though both people had departed. Although I have watched many nests of numerous species from blinds, I have never used this ruse but have entered in the dark before dawn or while the birds were absent, or, if one was in the nest, by stealing up unseen with the blind between us. I never thought birds too obtuse to distinguish one from two.

My belief that birds are not devoid of a number sense has been justified by numerous studies made since I began to use blinds in 1930. Early experiments on birds' ability to distinguish numbers were criticized on the score that they did not take sufficient care to avoid the "Clever Hans" type of error, so-called for the stallion that could tap out with his hoofs, according to a definite code, answers to difficult arithmetical problems. He deceived his trainer and other observers as to his ability to solve them until it was discovered that he could not do so unless the interrogator himself knew the answers. When the horse had tapped out the correct number, slight involuntary movements of the questioner revealed to the animal that it was time to stop—certainly a revelation of an acute intelligence, but not of the mathematical order. To avoid errors of this sort, O. Köhler, a foremost investigator of birds' ability to count, tested his subjects in a room where he and his coworkers could watch without themselves being seen by the birds.

In one test of a bird's ability to distinguish numbers, it was presented with a row of little boxes that had covers marked with different numbers of spots. Beside the boxes lay a card with as many spots as were on the lid that the bird should lift to reach a reward of food. A Common Raven and an African Gray Parrot learned to open the correct box from an array of them with, respectively, two, three, four, five, and six black spots on the lids. To make quite sure that the birds were responding to numbers rather than to some other feature of the experimental arrangement, the boxes were shifted at random, as was the distribution of the dots on their covers, and the shapes of the marks were varied; the only feature that remained constant was the correspondence between the key number on the index card the number of marks on the box with the reward. Despite all this confusing permutation, the raven, Jacob, consistently chose the proper box.

The foregoing was one of the experiments designed to demonstrate birds' ability to compare groups of units simultaneously visible. Another series of experiments was designed to test their ability to count a succession of objects or events, independently of any extraneous features that might be helpful. Thus, they were trained to eat only a certain number of grains from a big heap where random configuration offered no clue. Pigeons were taught to take only a particular number of peas that rolled into a cup singly and at randomly varied intervals of from one to sixty seconds, with nothing to distinguish the last permissible pea from the first forbidden one.

In another test of counting ability, a bird was required to take a certain number of food items from a long row of covered boxes arranged in different orders. The number of lids to be opened differed with each experiment. The test was made more complicated by changing the colors of the lids and the number of items that could be taken from each color. Undismayed, a Jackdaw learned to open black lids until he had taken two rewards, green lids for three, red lids for four, and white lids until he had eaten five items. This or another Jackdaw was given the task of taking five rewards from a row of boxes in which the first five contained, respectively, one, two, one, zero, and one. After he had raised the first three covers and eaten only four rewards, he went home to his cage, leaving his assignment unfinished. Presently he returned, went through the motions of taking one item from the first of the boxes that he had emptied, two from the second, and one from the third. Then he uncovered the fourth box, which had been empty since the beginning of the test, and passed to the fifth, from which he removed the fifth reward. Having accomplished his task,

he left the remaining boxes untouched and went home with an air of final-
ity. By bowing before each already opened box the same number of times
as he had already done, the Jackdaw showed that he remembered his previous activity, while by proceeding to the fifth he revealed awareness that
his task had not been finished. He behaved much as I might have done in
a similar situation.

From these and similar experiments, Köhler (as reported by W. H.
Thorpe) concluded that his birds were "thinking unnamed numbers," as
our ancestors doubtless did before they learned to count by a sequence of
named numbers. Tests have been devised to compare the competence of
birds and humans to recognize quantity without counting, as by projecting
a collection of objects on a screen, long enough for them to be clearly seen
but too briefly to be counted. Few people can recognize as many as eight;
pigeons can recognize five. The similarity of the abilities of humans and
birds in this respect suggests a common ground for the sense of numbers,
highly developed in humans only after we acquired adequate vocabularies.

In none of these tests by professional scientists were the numbers
named. It remained for an amateur, Len Howard, to demonstrate that a
bird could learn the names of numbers. Among the many unconfined
birds that freely entered and left her cottage for food or to roost was a female Great Tit named Star. Tits frequently tap or hammer to uncover
food, excavate nest cavities, or simply for amusement. One day it occurred
to their generous hostess to learn whether Star, an exceptionally energetic
bird, could count. When Star came to her hand for a nut, she withheld it
from the bird, saying, "You must tap for it." Looking directly at the tit, she
called sharply, "Tap, tap." While she spoke, the bird gazed intently at her
eyes. Then Star flew to the top of a wooden screen frame, a favorite perch,
and deliberately tapped twice with her beak, copying Howard's tempo,
after which Star flew to her hand for the nut. An hour later, Star responded correctly in exactly the same manner. Next morning, at the sug-
gestion of her instructress, she gave three resounding taps on the screen.

This was the beginning of a sporadic series of lessons extending over
about two years, in intervals when Star was not too busy defending her
territory, nesting, or engaged in other more exciting activities. She could
concentrate on arithmetic only when alone in the room; the presence of
other birds distracted her, even when they did not interfere with her les-
son by perching on the screen. Moreover, she was somewhat moody, turning her back to her teacher and standing with upraised head when not
inclined to tap. When she wanted a number, she rested on the screen with
her body tilted toward it and her bill almost touching it. At times she
became so absorbed in her lessons that she neglected to claim the nut that
rewarded a correct performance.

Great Tit

Star's instruction began with two taps on January 6. The number was gradually increased until by February 15 she repeated a series of eight, apparently her limit. Evidently she did not reach nine because Howard could not tap fast enough for the rapid mind of a bird; she would start to tap before her teacher reached the end of the long series. Star revealed her understanding of what she did by spontaneously dividing her taps into groups: two-two-two or three-three for six; four-four for eight; and in various other combinations. One day, when Star was given the number six, she tapped out five, then started toward Howard for her nut; but, realizing her mistake, she turned around and flew back to the screen, where she completed the number by tapping loudly once, then claimed her reward. This reminded Howard of the behavior of Köhler's Jackdaw when he uncovered only four baits instead of the required five.

After Star had become proficient in tapping any number from two to eight that was given to her in taps, Howard started to teach her to respond to spoken numbers by distinctly pronouncing "three" while tapping thrice, and so forth. The next step was to call the number without tapping. Again, Star proved to be an apt pupil. Before long she could tap out numbers from two to eight on verbal commands only. The depth of Star's understanding of numbers might have been investigated by giving her additional tests, such as picking up five nuts on hearing the word "five," but unfortunately this was not done. As far as I am aware, no bird has been taught to count beyond eight.

Star disappeared, probably victim of a cat, on April 22, 1953, when in her ninth year or older; the date of her birth was not known. Her most brilliant achievements were made in the last year of her life. Her history shows that simple procedures with free birds who have lost fear of humans can often teach us as much, if not more, about their mentality as do elaborate tests with captive subjects in the artificial situations of laboratories. Star's accomplishments appear to be due to her own superior intelligence no less than to her close rapport with an exceptionally sympathetic trainer. A few other Great Tits at the Howard cottage showed a slight inclination to tap out numbers but they never went far, perhaps for lack of motivation. It is regrettable that Howard did not take measures to avoid the Clever Hans effect while teaching Star to tap out numbers, thereby forestalling all doubts about the tit's remarkable achievement.

Because Great Tits are not mimics and are not known to pronounce human words, the best that Star could do was to tap out numbers. Some parrots have considerable mimetic skill and can learn to use human words correctly. Alex, an African Gray Parrot that Irene Pepperberg has carefully trained for many years (we shall meet Alex again in chapter sixteen), has revealed hitherto unsuspected mental capacities of a bird. Alex knows and correctly uses the English names of numbers from two to six. Not only can he tell the number of a homogeneous collection of up to six items, he can equally well declare the total number of diverse things presented to him on a tray. Likewise, he can tell the number of a certain kind of object in a mixed collection, such as the number of keys in an array of corks and keys.

In carefully controlled tests, the objects to be counted were variously or randomly arranged, to preclude the possibility that Alex was responding to a certain shape, such as a square or a pentagon, rather than to quantity itself. Alex's scores on these tests, ranging from 70 to 80 percent accuracy, left no doubt that he knew the meaning of the numerical labels he used. Whether Alex would learn to count as we do—saying aloud or to himself, "One, two, three . . ." as he successively touches or looks at a collection of objects, and realizing that the last number of the series represents the total quantity—has not been determined, according to the latest report of this continuing research available to me. P. Lögler had earlier demonstrated that Gray Parrots could solve problems involving numbers up to eight.

Of what use would the ability to count be to a bird? It seems safe to assume that large numbers would be useless to them, but small numbers may have more bearing. Possibly some birds count their eggs. To explain this matter, we must digress a little. Some species, called determinate layers, produce a definite number of eggs in a clutch; if eggs are removed while the birds are still laying, they do not replace the missing eggs, even if they do not abandon the nest; should an egg or two be added, they do not reduce the number they lay. The size of the clutch is physiologically fixed and not subject to change by what happens to the eggs.

The number of eggs that an indeterminate layer can produce in a sequence is not physiologically limited; while still laying, the birds can replace lost eggs in an effort to complete their clutch. The classic example of an indeterminate layer is a certain Northern Flicker whose eggs, except the first, were removed by a curious naturalist as they were laid, always leaving one in in the nest cavity. In seventy-three days, this prolific flicker laid seventy-one eggs in her frustrated attempt to achieve a normal set of five to nine. Also in the woodpecker family, a Wryneck, for which the usual set is seven to ten eggs, laid sixty-two when the eggs were successively removed. Northern House-Wrens, who lay from five to eight eggs if undisturbed, have produced up to twenty. By adding an egg or two to a Herring Gull's nest at the appropriate time, one may provoke her to lay only one or two eggs instead of the usual three. It is obvious in all these cases that some external stimulus, rather than purely internal factors, determines the number of eggs these birds lay. Whether the indeterminate layer controls her production by counting the eggs in her nest, or whether she continues to lay until the set simply looks right or feels right against her bare incubation patch, is not known. Knowledge of this detail might throw a little more light on the operations of a bird's mind. The subject has been reviewed in my *Parent Birds and Their Young* and more recently by E. D. Kennedy.

Counting chicks or fledglings, to make sure that none has strayed, might be of little use to birds. Mobile families keep together by voice as well as by sight. It would be unprofitable to neglect a brood while searching for one member that is probably irretrievably lost. Considering how rarely birds have occasion to count, it is remarkable that many of them can be taught to do so. Their ability to learn numbers points to a fundamental similarity of avian and human minds, despite the great differences. Apparently certain human groups could not, before contact with more advanced cultures, count beyond ten, the number of their fingers or toes. Is it more than a coincidence that birds, most of whom have eight toes, can learn to count to eight?

Counting and timing have much in common. Both are essentially the summation of units—objects in the first case; heartbeats, minutes, hours, days, months, years, or other intervals of duration in the second. However, there are important differences; counting books, coins, or other items is a conscious activity, which most educated people can perform accurately to a hundred or far beyond. Even humans with the best time sense, able to arise in the morning, go to work, retire, and so forth at fairly definite hours without constantly looking at their watches, do not continually count the minutes but rely upon subconscious processes. For organisms of many kinds, the basic regulator of activities is the circadian rhythm (from the Latin *circa* = about, and *dies* = day), so-called because it approximates the twenty-four-hour solar day although it may be slightly shorter or longer. Not only do many unconscious organic functions, such as slight daily fluctuations in blood pressure, conform to this rhythm but likewise the overt activities of animals are regulated by the alternation of day and night.

No avian activity is more important than the incubation of eggs, upon which the survival of each species depends. Successful incubation requires good timing. In probably the majority of songbirds and a number of other families, the female parent incubates alone, dividing her day into few or many sessions in the nest and recesses for foraging. These intervals may be short or long but they must be so related that the eggs receive sufficient warming. Most small birds, incubating alone, cover their eggs for about 60 to 80 percent of their active day, a few of the most assiduous for a little more; small flycatchers and birds with well-enclosed nests like those of becards for less time. It is surprising how one can watch two nests of the same species, separated by years and miles, and find that the birds attend their eggs with the same constancy in the same kind of weather. On cool days they may sit somewhat more constantly than on warm days, unless scarcity of food forces them to devote more time to foraging.

In many passerine birds and those of other orders, the sexes share incubation according to a schedule which ensures that the eggs will not be long neglected. Although the times of changeovers on the nest tend to be definite, when the incubation schedule repeats itself at daily intervals, a delay of an hour or so in the arrival of the off-duty partner is not likely to jeopardize the life or health of the sitting partner patiently awaiting relief, at least in normal circumstances when food is not scarce. But when, as in many marine birds, each partner sits continuously, fasting for days or weeks while its consort seeks food in the ocean, often at a great distance, an overlong delay in returning to the nest may be disastrous. Many breeding attempts fail because of improper timing. The on-duty partner may have been fasting so long that its nutritive reserves are becoming exhausted and it must abandon the egg(s) while it still has strength to reach the water and start foraging. If the off-duty parent does not return from the ocean with food about the time a nestling hatches, the hatchling may exhaust the nutrient reserves in its egg yolk and starve.

At the edge of the Antarctic continent, Adélie Penguins lay two eggs in an open nest outlined with small stones. During the incubation period of about thirty-four days, the parents cover their eggs in three sessions, the first and longest by the male, who has already been present, guarding the nest site while fasting, for more than two weeks before laying begins.

Adélie Penguins with chicks

After depositing her second egg, the female, who has spent nearly as long at the nest before laying, marches off to the ocean to forage. In about two weeks she returns, leaving her partner free to break his month-long fast. While she incubates, he takes a feeding excursion that tends to be shorter than hers and returns, if all goes well, shortly before an egg hatches. The whole success of the breeding enterprise depends upon the male's arrival in time to start feeding the newly hatched chicks.

To achieve this result, a competent father times his return not by the various lengths of the preceding sessions and foraging trips of the two partners but by the number of days since the eggs were laid. He appears to have an internal clock that sends him home thirty-four days, or a little less, after laying. Successful parents coordinate their movements to keep their eggs constantly covered and to ensure that one will be ready to feed the chicks when they hatch. If they fail to achieve this timing, the eggs are abandoned or the chicks starve. In a two-year study of Adélie Penguins, Lloyd Davis found that pairs who coordinated their incubation times and hatched their young were likely to reunite for the following breeding season; pairs who failed to coordinate more often separated.

Gray-faced Petrels were studied by Robin Johnstone and Lloyd Davis on Whale Island off the northwest coast of New Zealand, where these tube-noses lay single eggs in burrows in the ground. The exceptionally long incubation period of fifty-four days is covered by three major spells of sitting, each of about seventeen days, the first and third by the male, the second by the female, who also takes the final short session during which the egg hatches. As in the case of the male Adélie Penguin, the time of the female petrel's return to the burrow is determined not by the variable lengths of preceding foraging excursions but by the total interval since the egg was laid. An internal clock prompts her to return to her egg fifty-four days, or a little less, after she laid it, in time to feed her chick and prevent its starving.

Emperors, largest of penguins, incubate during the Antarctic winter, so that their slowly developing chicks will be ready to go to sea when the ice along the shore breaks up in the following summer, as recorded by Bernard Stonehouse. The female lays one big egg, delivers it to her partner, and walks away to forage while open water is still accessible. He places the egg on his feet, covers it with a flap of skin at the base of his abdomen, and stands with it on the ice or impacted snow with other incubating males, in the dusk and intense cold of the Antarctic winter. In fierce gales, they huddle tightly together to keep warm. When, after an incubation period of about sixty-three days, the chick hatches, its father, who has fasted for over two months, has only a little nourishment to regurgitate to it. The timely return of the mother, well fed after two months at sea, saves it from starving. Although wholly out of touch with the incubating male since her departure, she has somehow known when she would be needed.

It is improbable that the petrel or penguins have consciously counted the thirty-four, fifty-four, or sixty-three days between the laying of their eggs and the hatching of their chicks. Only a most exceptional human could accurately time such long intervals without a calendar, notching a tally-stick or some such device. Some subconscious process, physiological or mental, was evidently summing the days to prompt the birds to start homeward when the proper number had elapsed.

The circadian rhythm of an organism can be shifted by keeping it in a light-tight chamber on a schedule of light and darkness different from that of the solar day, as by periods of artificial lighting from midnight to noon. At Ithaca, New York, Judith Alexander and William Keeton did this with homing pigeons, who, like many other birds that travel by day, use the Sun as one of their means of orientation. When the pigeons' day has been advanced six hours by this treatment, they are carried in a covered basket to a point some fifty miles west of their home loft and released soon after sunrise. The poor birds are obviously in a predicament; the rising Sun is a little south of east when, anywhere north of the Tropic of Cancer, it should be in the south according to their advanced internal clock, which tells them it is noon. If they paid attention to the Sun's altitude in the sky, they would, despite their altered circadian rhythm, fly toward it to reach their loft in the east. But no! Disregarding its low altitude and giving attention only to its azimuth, or compass direction, they direct their courses as they should do if it were in the south at noon and fly with the rising Sun on their right, or northward. An advance of six hours in their circadian rhythm has shifted their course approximately ninety degrees counterclockwise, as retardation of the same amount would shift it clockwise. Some may eventually find their way home, others will be lost.

The behavior of these pigeons is remarkable. Placed in the same situation, a human would certainly be in a quandary; but I believe most of us would give greater credence to the direct evidence of our senses, distrust our altered time, and head toward the rising Sun in the east. Birds may be different; often they are more strongly influenced by what their whole experience, and that of their ancestors, tells them should be than by the immediate indications of their senses. This, I believe, is the explanation of some of the "stupid" things we sometimes see them do, as will be told in chapter twelve.

References: Alexander and Keeton 1974; Davis 1982, 1988; Howard 1956; Johnstone and Davis 1990; Kennedy 1991; Pepperberg 1990a; Skutch 1976; Stonehouse 1953; Thorpe 1956.

Chapter 7

🐚

Tool Using

TOOLS ARE OBJECTS other than parts of an animal's body that it employs for useful ends. They are extensions of its limbs, among which are the bills of birds, that increase their effectiveness or range. Toothbrushes, combs, knives and forks, needles, and pens are tools, no less than are brooms, saws, and shovels. Play things, which enhance the enjoyment of humans of all ages as well as of certain other animals, might by a liberal interpretation be included among tools, but for convenience they are best considered apart, as in chapter five. In a study of birds' intelligence, tool using is too important to be omitted. It reveals the flexibility of their minds, their ingenuity in solving problems. Although noted among widely differing species of birds, tool using is rare, or at least rarely observed and reported. As in our survey of play, so in our study of avian tool using we must depend largely upon incidental observations, supplemented by a few studies in controlled situations.

One of the best-known of tool users is the Woodpecker Finch, one of Darwin's finches of the Galápagos Islands. As described by David Lack and others, this plainly clad little bird climbs up and down upright trunks and over decaying branches, pecking into soft places and tearing off loose bark, eating the small invertebrates thereby exposed. When it detects a larval or mature insect too deeply embedded to be reached by its short bill, it cannot, like a true woodpecker, exsert a long, sticky tongue to extract the morsel but it has another recourse. It breaks off a long cactus spine or a straight twig from a shrub or tree where cacti are lacking. If the piece proves to be too long, short, or pliable, the bird drops it and seeks another. Then, holding its tool straight forward, in line with its bill, it pokes into the hole to draw or drive out the insect, drops its probe, and devours the prey. Sometimes the finch carries a spine or twig from tree to tree, poking

Woodpecker Finch

into holes and crevices. Similar behavior has been reported of the Mangrove Finch on a different island of the Galápagos Archipelago.

Woodpecker Finches were studied in an aviary by George Millikan and Robert Bowman. There the birds used toothpicks, pine needles, lengths of thin wire, even a long stick of spaghetti, to dislodge mealworms from narrow slits in a board or from other crannies. If the piece proved to be too long, they tried to break it or they might hold it by the middle instead of the end, thereby shortening its effective length. As in the wild, they did not pick up tools from the ground. Often the finch held its tool beneath a foot while it ate. These investigators tried to teach tool using to other Galápagos finches by placing them in view of Woodpecker Finches. Only one of them, a Large Cactus Ground-Finch, picked up sticks and stuck them into cracks, apparently having learned the trick from its neighbors—a highly significant observation. On Santa Cruz Island of the Galápagos, Margaret Hundley watched a Warbler Finch carry a three-inch petiole or flower stalk and try about four times to probe into a crevice in bark, but the piece was too flexible to enter and the little bird tossed it away.

Using a twig to extract insects from crevices has rarely been seen in birds of other families. In Texas, Douglas Gayou watched a Green Jay take a small twig in its bill and poke around beneath loose bark on a dead branch until it removed the probe with an insect attached. Holding the twig beneath a foot, it ate the creature. A little later, the jay pried off a flake of bark with another twig and consumed the exposed insects. A juvenile took a twig and tried to procure food in the same manner as its parent but was unsuccessful. Of fourteen Green Jays whom Gayou watched while they foraged, only these two used tools—behavior apparently otherwise unknown in wild jays. However, Ronald Orenstein watched a bird of the same family, a single New Caledonian Crow, on the Pacific island of that name, carrying a twig that it stuck repeatedly under bark and into the ends of hollow branches, apparently without dislodging anything edible.

Similar use of tools is described for other species under "Tools, use of" by J. H. R. Boswall in Campbell and Lack's *Dictionary of Birds*. Several free Orange-winged Sitellas, Australia's counterpart of nuthatches, held in their bills splinters of wood that they poked into cavities to remove insect larvae, which they ate while they held their tools under their feet. In Tanzania, a Gray Flycatcher used a grass stem to extract winged termites from crevices in a concrete veranda. In the London Zoo, an Oystercatcher regularly used a stick to probe fissures in the concrete walls of its enclosure, occasionally driving out an insect. In a different context, an American Robin raked aside fallen leaves with a stick.

In the longleaf pine forests of Louisiana, Douglass Morse watched Brown-headed Nuthatches employ readily detached scales of bark to remove less readily detached bark from trunks and large limbs of the pine trees. Holding the tool in its bill, a bird inserted the bark under a loose scale and maneuvered it as a wedge or a lever to pry the other piece off. When it succeeded, it dropped the tool and detached piece to the ground and devoured whatever edible was disclosed. Occasionally, a nuthatch tore off three or four scales of bark before discarding its tool. In 150 hours of observation when the pine seeds that they crack and eat were scarce, Morse saw Brown-headed Nuthatches use tools ten times; when this preferred food was abundant, the birds pried off bark only once in seventy-five hours. In forests of loblolly pine and spruce pine, the bark of which is not so easily detached from the trees as is that of longleaf pine, he did not see the nuthatches try to remove pieces of it.

In Ecuador, Carl S. Berg watched a Pale-mandibled Aracari try vainly to reach with its long bill an insect that had taken refuge in a deep cavity in a decaying branch. Changing its procedure, the bird broke off a neighboring twig about eight inches (20 cm) long. Holding the twig beneath a foot, the aracari removed the twiglets by repeatedly pulling the twig through its serrated bill, until it had a straight, clean branch, with which it probed the hole where the intended prey lurked. When clumsy repetitions of this procedure failed to yield an insect, the aracari resumed its more profitable fruit eating.

Instead of using tools to disclose or extract hidden food, some birds draw exposed items to within reach by strings or other means. W. H. Thorpe describes how, at bird feeders in England, food is suspended by threads for free Great Tits and Blue Tits to reach. Without fumbling, a tit standing at the edge of the table hauls up with its bill the thread from which the titbit dangles, successively holding lengths of thread beneath a foot until the prize comes within reach. One of the most persevering Great Tits pulled up a string two feet (64 cm) long. Among other European birds who have learned to procure food by string pulling are Coal Tits, Redpolls, Siskins, and Jays, and among exotic birds Budgerigars and Blue-and-Yellow Macaws.

European Goldfinches so readily learn this trick that for centuries they have been kept in cages so designed that they must earn their food by pulling up and holding tight with their feet a string attached to a tiny cart resting upon an incline, and their water by hauling up a thimble containing it. In the sixteenth century, goldfinches were so widely kept in this fashion that they were known as "draw-water" or its equivalent in several European languages. Across the Atlantic, Millikan and Bowman tested the string-pulling aptitude of a number of species. Five captive Darwin's finches, including the Woodpecker Finch, learned to reach food attached to strings, but Northern Mockingbirds, Redwinged Blackbirds, Brewer's Blackbirds, White-crowned Sparrows, and Cuban Grassquits failed. Of the mainland birds, only the Plain Titmouse was successful.

In Brazil, captive Blue-black Grassquits and seedeaters of the genus *Sporophila* have learned to pull up small vessels of water dangling beneath their perches. Helmut Sick, to whom we owe this information, remarked that the movements needed to gather in the string that holds a receptacle do not differ much from the ones these little birds use to pull a stem of grass through a foot while they pluck off the seeds.

As far as I know, string pulling has been witnessed only once in a wholly natural situation. In North Carolina, J. C. Dickinson, Jr., noticed a caterpillar hanging by a silken thread about eighteen inches (46 cm) long, beneath a branch of a tree in which a mixed flock of small birds foraged. A Tufted Titmouse leaned over the limb, grasped the thread in its bill, and hauled up a length that it held beneath a foot while it drew up another length, continuing this until it reached and ate the caterpillar. Caterpillars of various kinds dangle in the air on threads that they secrete. Hanging so exposed, they offer meals to birds, who might either catch them in flight or draw them within reach as the titmouse did. This may be a natural use of string pulling but it appears to have been recorded rarely in the wild.

In many parts of Norway and Sweden, Hooded Crows make a very different use of string pulling. In early spring, they draw out anglers' lines set through holes in the ice, to steal the fish or bait. Grasping the cord in its bill, the crow walks slowly backward with the line as far as possible. Then it returns to the hole by walking forward with its feet on the line, thereby preventing it from slipping back into the water. If necessary, the bird repeats this sequence until the fish or bait comes within reach, as described by J. C. Welty.

Birds show great ingenuity in reaching food difficult of access. T. B. Jones and A. C. Kamil reported that an adult, hand-reared Blue Jay tore a long piece of newspaper from the floor of its cage, then pushed this strip between the wires of an adjoining cage and shoved the paper around until it brought pellets of food within reach of its bill. Acting on this hint, the jay's keepers gave it a paper clip, a straw, and a tie for a plastic bag, all of which the bird maneuvered to rake pellets from the adjacent cage close enough to eat them. Blue Jays were not known to use tools in the wild.

Resting beside a pond or smoothly flowing stream, one often sees minnows approach to investigate a berry, a bit of twig, or other small object that falls from overhanging boughs or that one throws upon the surface. Apparently from observing this habit of fishes, Green-backed Herons have learned to bait them by casting edible or inedible things upon the water. On Lake Eola at Orlando, Florida, where water birds liberally fed by visitors had become tame, Harvey Lovell threw a piece of bread to a Green-backed Heron fishing beside the lake. The bird picked it up and placed it in the water, not to soften it but to attract fishes, one of which it speared while the fish nibbled at the bait. Other foods were used for the same purpose. This was apparently the first report of a habit widespread in the almost cosmopolitan Green-backed Heron. Subsequently, it has been seen to bait fishes with bread, fish-food pellets, mayflies, and feathers, all in North America, by C. R. Preston and associates.

In Japan, Hiroyoshi Higuchi watched Green-backed Herons fish with flies, cicadas, grasshoppers, insect larvae, earthworms, leaves, bits of bark, moss, feathers, and diverse pieces of rubbish, as well as crackers people brought for the fishes. The herons held twigs beneath a foot while they broke off short lengths to cast upon the water—an example of tool making. Nearly all these lures brought some victims within reach of the herons' beaks but flies were by far the most attractive, with other adult insects next in order. By using baits, herons greatly increased their rate of catching prey. The art of baiting fishes had to be learned. Juvenile Green-backed Herons often tried but succeeded only with flies, and that rarely. They often used oversized baits and they failed to crouch and make themselves less conspicuous while they waited for fishes to approach. In Kenya, Pied Kingfishers lured fishes within range by repeatedly offering them bread.

Birds use tools not only to procure food but also to prepare it for eating after it has been found. Egyptian Vultures enjoy the full meal that a huge Ostrich egg provides, but the shell is too hard to crack with their bills. To overcome the difficulty, this vulture picks up small stones in its

Egyptian Vulture

bill and, stretching its head high, forcefully hurls them at an egg. Their aim is far from perfect—the most skillful hit the egg on only 40 to 60 percent of their throws—but they persist for many minutes until they strike the target. Stone throwing is not, as some have conjectured, learned by watching experienced adults but innate, as C. R. Thouless and his coworkers proved in experiments with hand-reared vultures that lacked teachers. The widespread habit of casting stones at Ostrich eggs appears to be derived from the innate behavior of hurling large eggs, such as those of the Great White Pelican or the Lesser Flamingo, against a stone anvil or other hard surface, as the easiest way to open them. Vultures prefer roundish or egg-shaped stones, weighing about one and a half ounces (46 g) for hurling against Ostrich eggs. Although throwing is an innate behavior, the vultures need to learn that an Ostrich egg is a source of food, as they apparently do by finding some in shells already cracked by their parents or otherwise. Reports that in Australia Black-breasted Buzzards crack Emus' eggs by dropping stones on them from a height were accepted as valid by A. H. Chisholm but have been doubted by other naturalists.

Recently, Jeffrey S. Marks and C. Scott Hall reported that Bristle-thighed Curlews wintering in the Hawaiian Archipelago on Laysan and Tern islands pick up fragments of coral less than one inch in diameter and slam them against the big eggs of Laysan and Black-footed albatrosses, eggs with shells too thick to be pierced by the curlew's sensitive bill-tip. Through the perforation thus made, the bird extracts the egg's contents. This behavior is related to the curlew's habit of hurling food items too big to be swallowed whole, such as crabs, to the ground after raising them above its head. When necessary, they repeat this maneuver until the article is reduced to swallowable size. The behavior is evidently innate. Flightless chicks on the breeding ground in Alaska slam inedible objects, such as lichens and mosses, and hatching-year curlews newly arrived on the Pacific islands do the same with seabird feathers, old crab shells, and seaweeds. Adults slam only edible objects but they have not been seen doing this while rearing their broods in Alaska.

In *The Malay Peninsula,* Alfred Russel Wallace told in detail how the big black Palm Cockatoo breaks open the kanara nut, a favorite food but with a shell so hard that only a heavy hammer will crack it. Taking a nut endwise in its great bill and holding it firmly by pressure of the tongue, the bird cuts a transverse notch by a sawing motion of its sharp-edged lower mandible. This done, it holds the nut in a foot while it tears off a piece of leaf, which it retains in the deep notch of the upper mandible. The elastic tissue of the leaf prevents the nut from slipping, while by a powerful bite the bird breaks off a piece of shell. Again taking the nut in a foot, the cockatoo now uses the long, sharp point of its upper mandible to detach bits of the kernel and removes them with its long, extensible tongue.

The habit widespread in gulls of breaking open the shells of mollusks by dropping them from a height upon rocks or other hard surfaces is often considered an example of tool using. However, because the gulls do not hold and manipulate the rocks, they are not properly its tools, but they may be viewed as substitutes for tools. Benjamin Beck spent many hours studying the behavior of Herring Gulls on Cape Cod, Massachusetts. These gulls had the choice of two surfaces on which to drop their prey: a long, narrow, concrete seawall and two broad, paved parking lots. These lots were targets too wide to miss but the seawall required greater accuracy, especially in the prevailing strong winds. Only 68.5 percent of the shellfish dropped over the wall actually struck it, while almost all that were taken to the parking lots fell upon them. The gulls adjusted to this problem by dropping from a lower altitude over the wall than over the lots, where the longer and harder fall more frequently smashed the hard shells.

Less often than to procure food, birds use tools to care for their plumage. In the Florida Keys, Andrew Meyerriecks watched a Double-crested Cormorant squeeze preen oil from the uropygial gland above the base of its tail and, with broad sweeps or dabs of its anointed bill, apply the secretion to its wing feathers. Suddenly, while the bird spread its wings, a feather loosened by molt was blown away by a breeze and landed nearby. After staring at the plume for a few seconds, the cormorant picked it up, held it crosswise in the bill, then adjusted its grasp until the feather projected straight forward, its apex outward. With this brush, the cormorant removed more oil from the gland and spread it over both wings. In a different context, two captive Lesser Sulphur-crested Cockatoos and two African Gray Parrots held objects in their feet to scratch their backs. G. A. Smith, who described this behavior, suggested that the parrots scratched themselves because they lacked companions who, in normal circumstances, would have preened them.

Probably the widespread, puzzling behavior of "anting" might be regarded as a form of tool using, usually with living tools. Nearly always on the ground in the north temperate zone, mostly in a tree or shrub in tropical America, a bird seizes in its bill an ant or some substitute, such as a camphor ball or a pungent fruit. Assuming a characteristic posture, the bird sweeps the object over the inner surface of its half-spread wings or the underside of its forwardly bent tail, anointing these feathers with formic acid or other secretions, after which it drops or swallows the insect. Whether the purpose of this activity is to control feather parasites or to stimulate the bird in some way is not clear.

Occasionally, tools are used in connection with nests or bowers. Often included among tool users is the Indian Tailorbird, an Old World warbler, who perforates the edges of a living leaf with its bill and, passing cobweb through the holes, binds the edges together to make a cocoon for its nest. Why the tailorbird should be included in the category of tool users, but not hermit hummingbirds, who with liberal applications of cobweb fasten their nests beneath the tip of a palm frond or other leaf, is not immediately clear. The difference appears to be that the tailorbird draws the edges of a leaf together with silken strands, in which case the strands might be regarded as tools, whereas the hummingbird uses the leaf-tip as she finds it; the cobweb is only part of her building materials. Several small African warblers also bind leaves together to cover and conceal their nests and might be included among tool users. In a zoo, an Eclectus Parrot used part of a palm leaf to scratch out a hollow for her eggs; but in the wild these parrots occupy holes high in great trees. Bills are so adequate for building even the most elaborate of nests that birds hardly ever need the aid of tools.

The Satin Bowerbird of Australia and several other bower builders paint the inner walls of their avenuelike constructions with the pulp of berries, green liverworts, or laundry blue where available, mashed or ground in their bills and mixed with saliva to form a paste, as described by A. J. Marshall. While applying this preparation, they hold in the mouth a spongelike wad of fibrous bark or some similar material, apparently more to keep the stuff from dribbling out than to use it as a paintbrush. Nevertheless, the wad conforms to our definition of a tool.

S. G. and M. A. Pruett-Jones told how, in montane rain forests of Papua New Guinea, a Lawes' Six-wired Bird of Paradise prepares a stage for courtship by carrying all movable leaves, twigs, and other litter from a small patch of ground, leaving it bare as though swept with a broom. A foot or two above this court, a thin, more or less horizontal branch or vine serves as the bird's display perch. Holding in his bill a fragment of cast snakeskin, a small piece of chalk, or a tuft of fur or feathers, the bird rubs the object methodically over his perch, sometimes continuing for many minutes, until the perch becomes smooth and shiny, and streaked with white when fragments of chalk from a calcareous outcrop are included among his tools.

A curious example of tool using by White-breasted Nuthatches was described by Lawrence Kilham. Holding an insect or fragments of crushed, juicy-looking vegetation in its bill, a nuthatch of either sex sweeps the nest box or hole in a tree where it breeds, inside, outside, and over the surrounding surface, often continuing this activity vigorously for many minutes. Frequently, it rubs over the nest a metallic blue blister beetle, which secretes from its leg joints a copious, oily, blistering fluid. The presence of a squirrel in a neighboring tree often incites intense sweeping. By surrounding the nest, especially the doorway, with a pungent or repellent substance, nuthatches try to hold aloof flying squirrels, which covet nest boxes and nest holes for shelter, and other squirrels that eat eggs and nestlings. Only one other bird is known to bill-sweep. In the Chiricahua Mountains of Arizona, Millicent and Robert Ficken watched a Mexican Chickadee repeatedly anoint the area below the entrance to its nest with the juices of small insects that appeared to be beetles.

A few birds use offensive or defensive tools, commonly called weapons. Australian Brush Turkeys kick barrages of gravel and litter at monitor lizards that compete with them for food. In Israel, an Egyptian Vulture killed a monitor lizard with a stone. In New Jersey, a Fish Crow, carrying in its bill a dried stalk of marsh grass, flew directly toward a Laughing Gull incubating its eggs. The crow hovered about fifteen feet above the gull, then transferred the stalk to its feet and descended lower over the gull, who became increasingly agitated. The crow dropped the grass above the gull but a breeze carried it aside; had it struck or landed near the gull, the bird would probably have fled, exposing its eggs for the aggressor to devour. As it happened, the crow flew away, leaving the nest unharmed. On an island off the coast of Newfoundland, a Common Raven, perching about a yard above a Black-legged Kittiwake sitting on a nest on a cliff, called loudly and alarmed the sitting bird. The raven pulled a tuft of dry grass from the face of the cliff and dropped it on the white gull, making the kittiwake flee, whereupon the aggressor rummaged through the nest, finding it empty. W. A. Montevecchi, to whom we owe these observations, often saw ravens drive kittiwakes from their nests by approaching them with raucous calls or attacking them.

When Stewart Janes climbed to a Common Raven's nest on the side of a high cliff, one of the parents tossed seven stones, up to three inches long, toward him from the cliff-top. One of these missiles struck his leg. On his second visit later that same day, a raven threw gravel at him, apparently because no more stones were readily available.

A few miscellaneous observations, from *A Dictionary of Birds,* on captive birds may be added to the foregoing accounts of tool using. An American Crow filled a plastic cup by dipping it into a water trough and carried it fifteen feet up to moisten its dry mash. Several parrots have used hollow objects to dip up water for drinking.

Perhaps to include in a chapter on tool using the behavior of certain gulls and terns that bring stones to their nests is not too far-fetched. Some of these birds, including Herring, California, and Ring-billed gulls and Common Terns usually lay clutches of three eggs and have three bare incubation patches to warm them. When, from failure to lay the third egg or having lost it, they have only two or one, they are ill at ease and try to correct the deficiency by rolling into the nest small stones from nearby. As illustrated by Michael R. Conover, these stones are rounded and about the size of the birds' eggs; sometimes the resemblance is remarkably close. Malcolm C. Coulter learned that gulls with three eggs sit for longer spells, and less restlessly, than those with more or less and that their incubation period is correspondingly shorter. Apparently, the birds rest more comfortably with smooth objects in contact with all three of their bare incubation patches, and the substitution of an appropriately sized foreign object for a missing egg increases the efficiency of incubation. Whether these larids mistake the stones for eggs is not clear. Some can distinguish their own eggs from those of other members of their species, making it improbable that they would be deceived by stones. On the other hand, some gulls will continue to sit upon the rocks after all their eggs have been removed.

Our examples of tool using by birds are few but diverse. Many are unique. The cormorant anointing its plumage with a detached feather, the Green Jay and New Caledonian Crow using probes to dislodge insects, and the Blue Jay pulling food pellets within reach of its cage with a strip of paper are the only individuals of their species known to behave as they did. A more thorough search through ornithological writings than I have made would surely disclose additional instances of tool using, but I am certain that they would involve only a tiny fraction of the approximately nine thousand species of birds. In nearly seventy years of bird-watching, mostly amid rich tropical avifaunas, I have never seen a bird with a tool except when anting. Birds so rarely use tools because they have little need of them, being superbly equipped with organs adequate for all their necessary activities. Like an elephant's trunk, a bird's bill is, in effect, a fifth limb, a versatile organ capable of performing many diverse tasks: gathering food, preening plumage, building nests for breeding and sleeping, attending its young, defending them from animals not too big and powerful.

Compared with the neat structures that birds make, the platforms that gorillas and other apes pull together for sleeping are crude affairs. Among mammals, only a few rodents build as well as birds do; the globular nests of the dormouse compare favorably with those of many wrens and among larger creatures, beavers' lodges are not inferior to the covered nests of Hamerkops, or Hammerheads. Indeed, I venture to assert that, except for lifting, carrying, and fighting, the hands of hominids could accomplish little more than do the bills of birds until, about two million years ago, our ancestors began to make cutting tools by chipping flint or obsidian with other tools, especially flints. Not until, only a few thousand years ago, humans developed metallurgy were we able to make instruments and machinery that placed us far ahead of birds in constructive ability.

Since tool using by birds is rare, why, it may be asked, does it merit consideration in a book about their minds? It is precisely the rarity of tool using by birds that makes it so instructive; it reveals the flexibility of their minds. Most of the things that we see birds do are programmed in their genes and conform to hereditary patterns, although their performance is usually improved by practice and learning. But occasionally a bird solves a problem or finds a better way to perform a habitual activity. An excellent example of this is the cormorant using a feather to anoint its plumage. It was reaching its preen gland with its bill in what appeared to the observer an "awkward manner" when the sight of its molted feather suggested, almost immediately, a better way. In a flash of insight, the bird picked up the feather and proceeded to use it as, perhaps, it had never done before. Similarly, the captive Blue Jay who tore off a piece of newspaper to pull food pellets within reach acted in a way that an unconfined jay would have little occasion to practice. In its boring confinement, the bird suddenly had a bright idea! Likewise, the American Crow who carried up water to make its meal of mash more palatable was capable of making an innovation in the behavioral repertoire of crows. In each of these episodes, all the elements of the situation were immediately present to the bird. Such spontaneous acts, so rarely observed, are often more revealing of how a bird's mind operates than is learning by laborious repetition to solve the artificial problems prepared for birds in researchers' laboratories, although these, too, add to our understanding of birds' minds.

In insight learning, a bird sees a need or a problem and solves it by combining its capabilities in a new pattern of coordinated movements, as is well illustrated by string pulling to obtain food. Many birds have frequent occasions to tug or pull, as in detaching berries from twigs or gathering materials for nests. Many hold food beneath a foot while they tear pieces from it; many others are not known to use their feet for holding. Birds accustomed to holding with their feet, such as titmice, can often learn string pulling; they need only to see the problem whole and use skills they already have. I surmise that vireos, and especially the related pepper-shrikes, who frequently hold with their feet, could readily learn string pulling; but I doubt that tanagers and wood warblers, whom I have never seen use their feet in this manner, could easily learn.

Did a habit so well established in a species as probing with spines and twigs by the Woodpecker Finch originate as a genetic mutation, or a series of them, or as the invention of a clever individual long ago? We do not know; we do not even know whether it is innate or must be learned by each individual finch, as might be determined by appropriate experiments with hand-raised birds. That a rewarding innovation by some enterprising bird may spread rapidly through a population is evident from the example of the Great Tits who steal milk from householders' bottles delivered on the doorstep in England. Such an innovation, perpetuated for generations by example or tradition, may finally be made innate by appropriate mutations, the process known as genetic assimilation, which simulates the direct inheritance of acquired characters but depends on favorable random mutations.

To use tools, birds need no obvious structural changes; they simply pick up the tool in a bill that is adapted for other ends. Tool using is a supplementary rather than the main mode of foraging of the birds in which it has been reported. For their principal method of foraging, many birds exhibit profound anatomical modifications. Think of the long, slender bill and tongue and wings specialized for motionless hovering of a hummingbird; the chisel-like bill, protrusile tongue, and grappling toes of a woodpecker; or, to turn to a mammal, the long neck of a giraffe, with its specialized circulatory system. How did all the anatomical and behavioral idiosyncracies of these animals arise? Did the structural changes precede or follow the behavioral changes? Did hummingbirds never visit flowers until they were endowed with a perfected method of extracting nectar from them? Did giraffes delay browsing on high branches until they had evolved exceptionally long necks?

The diverse anatomical peculiarities of a specialized forager could hardly have arisen at a stroke by a single mutation; they required a series of mutually supportive mutations spread over many generations. Almost certainly, the habit preceded the anatomical specializations that made it highly profitable. As competition for low foliage increased, certain ancestral giraffes stretched up to the limit of their short necks to reach high leafy boughs. As this habit spread through the population, random mutations that increased the neck's length gained immediate survival value. Similar mutations might occur in a quadruped that grazed or browsed low but would by natural selection be eliminated as valueless or deleterious. By the accumulation of complementary mutations over the generations, giraffes acquired the ability to browse on foliage far above the reach of antelopes and other quadrupeds that shared the African savannas with them. Likewise, ancestral hummingbirds, with bills perhaps a little longer than those of associated birds, occasionally reached nectar in insect-pollinated flowers, giving value to mutations that increased the length of their bills and tongues and their ability to hover motionless while they drank. Similar mutations in a bird that never tried to sip nectar would be valueless and discarded.

Habits begun by individuals with minds alert to exploit new sources of food, or to gain other advantages, appear to have been precursors of many morphological changes. In certain important aspects of evolution, mind has led the way. If mind had not profoundly influenced the course of evolution, evolution could not have improved the neurological foundations of intelligence.

References: Beck 1986; Berg, letter; Campbell and Lack 1985; Chisholm 1954; Conover 1985; Coulter 1980; Dickinson 1969; Ficken and Ficken 1987; Gayou 1982; Higuchi 1986, 1988; Hundley 1963; Janes 1976; Jones and Kamil 1973; Kilham 1971; Lack 1961; Lovell 1958; Marks and Hall 1992; Marshall 1954; Meyerriecks 1972; Millikan and Bowman 1967; Montevecchi 1978; Morse 1968; Orenstein 1972; Preston et al. 1986; Pruett-Jones and Pruett-Jones 1988; Sick 1993; Smith 1971; Thorpe 1956; Thouless et al. 1989; Wallace 1872; Welty 1975.

Chapter 8

Aesthetic Sense

IN LIGHT second-growth woods, tall thickets, and at the edges of old forest, here in southern Costa Rica, I sometimes find roundish patches of ground, a foot or two (30 or 60 cm) in diameter, that appear to have been swept clean with a broom. A careful search may disclose from one to a dozen similar patches scattered nearby but a group of these courts may be far from any others. Beside each bare patch grow two or more slender saplings. Dropping a leaf or flower on the cleaned ground of one, I retire to watch from a distance. Presently a stout little bird with a broad orange collar and black crown arrives, spies the intruded object, drops to the ground, seizes it in his short bill, and carries it away. Continued study leaves no doubt that each bare patch was made by a single male Orange-collared Manakin for his private use. The grouped courts form a courtship assembly, or lek, in which the males gather, year after year, to compete with one another for the privilege of mating with the females that they cooperate to attract.

After removing the leaf or flower, the court's owner may return to display. He leaps in a low arc, back and forth over the court, from sapling to sapling on opposite sides. Each brisk jump is accompanied by a loud *snap*, like the sound of breaking a dry twig, made by his highly modified wing feathers. If I wait patiently, I may witness the arrival of an olive-green female, who resembles the elegant male only in her size and orange legs. The two jump back and forth, in opposite directions, passing each other in the air, while he snaps and she leaps silently.

This performance alerts other males. From the surrounding undergrowth I hear them snapping as each tries to entice the demure visitor to his own court; he will not intrude upon another's. More often than not, after jumping with one suitor, the female abruptly departs to perform with another. Free to choose, she apparently selects the father of her nestlings by differences too subtle for me to detect. If I am exceptionally fortunate,

I witness the culmination of one of these jumping bouts. After a final leap over the court, the female clings to one of the upright stems beside it. Her partner alights above her, and slides down to her back. After the conclusion of their brief union, she goes off alone, probably to finish the little cup in which she will lay two eggs, to hatch them and rear the nestlings all alone.

The Orange-collared is one of about sixty species of manakins, a passerine family most closely related to the cotingas and American flycatchers, that inhabit tropical America from southern Mexico to northern Argentina, mostly in warm lowlands. In a few species, the males are as plainly attired as the females; but many are elegant little birds with varied and striking color patterns. Their courtship displays are as diverse as is their coloration. Only the four species of the genus *Manacus* to which the Orange-collared belongs are known to make bare courts; others perform in trees. In several species, two or three males join in elaborate, coordinated displays before an intently watching female, after which she briefly mates with one. These fascinating birds introduce us to far-reaching questions. Why are the males so much more ornate than the females? What determines the female's choice of a partner? Have birds an aesthetic sense?

Some of the features that make birds beautiful can be explained by widely applicable evolutionary principles. The plumage that keeps them warm and streamlines them for flight gives them graceful forms. Feathers, pleasing in texture, are an excellent base for the colors that embellish birds. The green of parrots makes them difficult to detect while they rest quietly in the verdant crowns of trees. The widespread olives and browns make their wearers more obscure in the deep shade of forests. The mottled plumage of nightjars blends with the dark, sun-flecked litter where they rest by day. The colors of aquatic birds, dark above and pale below, make them less conspicuous when viewed from above, against the water, or from below, against the sky as background. Bright colors are appropriate for sunny treetops.

Although the plumage and coloration of birds are, on the whole, utilitarian, some of their adornments are so brilliant that they appear gratuitously to attract predators. The lavish ornamental plumes or wattles of many males not only appear burdensome impediments to flight and foraging but unfit them for attending nests. The peacock's gorgeous train, the cascading plumes of some birds of paradise, the hundreds of realistic eyespots on the Great Argus Pheasant's wings that so fascinated Darwin, the same bird's excessively long tail, the dangling wattles of bellbirds and umbrellabirds, and many other apparently superfluous adornments— what do they contribute to the survival of their bearers? These embellishments seemed so incompatible with Darwin's views of natural selection as

Orange-collared Manakin (male)

the shaper of evolution that he sought a supplementary principle and, in a book published in 1871, proposed the theory of sexual selection, which continues to be examined and revised by contemporary biologists.

These two modes of selection differ profoundly. Sexual selection is the choice of partners for the perpetuation of life; natural selection is largely the extermination of the less fit by sundry agents of death. Sexual selection follows the methods of the intelligent breeder of animals and plants; natural selection follows the ways of predator and plague. Although sexual selection is sometimes called a mode of natural selection, it is this only because everything that occurs in nature is natural. Natural selection would eliminate the extravagant ornaments, costly of materials, that sexual selection often produces. Sexual selection provides our earliest examples of the selection by one animal of another for its personal qualities rather than simply as a member of a species. It is the forerunner of individuality.

We now distinguish two main categories of sexual selection, intrasexual and intersexual, or within a sex and between sexes (table 1). The former recognizes the competition among individuals of the same sex for access to the other sex, usually by males for females. Intrasexual selection is especially intense in harem-forming mammals, such as some seals and ungulates; among birds it is exemplified by domestic fowls and certain other members of the pheasant family, by Red-winged Blackbirds of the marshes, some other icterids, and a few less familiar groups. It is responsible for the larger size of males than females and for the great development of fangs, horns, spurs, and other weapons used in often fierce and

Table I. Types of Sexual Selection

Type	I Intrasexual	II Unilateral Intersexual[a]	III Mutual Intersexual
Characteristic	The "law of battle"	The contest to charm	The quest of a partner
Quality proximately selected	Belligerence	Conspicuousness and beauty	Compatibility
Activity of males	Fighting and hostile display	Visual and/or auditory display; posturing, dancing, calling	Visual and/or vocal display; often obscure
Activity of females	Mostly passive acceptance	Active choice	Sometimes same as male's
Duration of union	Until after coition	For coition only	Prolonged, often lifelong
Effects on male	Development of weapons, often also of exaggerated size	Development of bright colors and ornaments, vocal and/or instrumental sounds, and/or bizarre antics	Involvement in family care, sometimes increase in brilliance and song
Effects on female	Sometimes acquires male's weapons by genetic transference	Usually remains dull and self-effacing	Tends to resemble male in appearance and voice
Effect on reproductive rate	Probably slight	In birds, limits size of brood	Permits raising of more young
Quality ultimately selected	Strength and vigor	Vigor	Cooperativeness, constancy

SOURCE: From Skutch 1992. Reprinted by permission of University of Texas Press.

[a]From one point of view, this is also a mode of intrasexual selection, since the males compete with one another and may exclude one another from mating. But the distinctive feature of this mating system is free choice by the female, hence it is more appropriately called "intersexual selection."

sanguinary struggles among competing males. Governed by "the law of battle," intrasexual selection promotes belligerence more than beauty and need not detain us longer here.

Intersexual selection is the free choice of partners in reproduction. It may be either unilateral, "the contest to charm," when males compete to attract females, as in manakins, or mutual, "the quest of a partner," when a pair is formed by a male and a female who attract each other. Mutual selection occurs chiefly among permanently resident birds, who often unite, soon after they become independent of their parents and long before they will breed, in pairs that persist through the year and often as long as both partners live. It tends to make the partners alike and often beautiful. It appears to be due to mutual selection that both sexes of tropical species tend to be equally colorful, even in families such as the tanagers, wood warblers, and orioles—families in which the females of northern, migratory species are frequently much plainer than the brilliant males.

Intersexual selection prevails in migratory or nomadic birds of which males arrive on the breeding grounds before females, claim territories, and sing profusely to attract mates, as in many wood warblers, Old World warblers, thrushes, finches, and numerous other familiar songbirds of northern lands. In these species, a female's choice is influenced not only by the appearance and/or song of the male but also strongly by the resources of the territory he offers. The most striking effects of intersexual selection are achieved by males who offer only paternity to the females—not territory, nest sites, protection, or help in rearing the young—as in manakins. Several males often gather in a fairly compact courtship assembly, the lek, in which each participant holds a small individual territory, or court, where he performs. In some species the males are more widely separated, within hearing but not within view of one another, in what is called an exploded or dispersed lek. Or the displaying males may be quite solitary but probably accessible to several females. Year after year, as long as the habitat remains favorable, the birds gather in the same place, which becomes well-known in the neighborhood and, with the sounds of the displaying males, is readily found by females with developing eggs that need to be fertilized. These females are usually cryptically colored and rather silent, in strong contrast to the brilliant and often vociferous males that they seek. Exceptional are a number of species of hummingbirds that court in assemblies despite having little or no sexual differences in plumage.

Among birds that court in assemblies are many of Earth's most beautiful, including many hummingbirds, manakins, birds of paradise, the Ruff in the sandpiper family, and some grouse. As one would expect of such skillful fliers, hummingbirds display on the wing, but most others perform on trees, shrubs, or the ground. Their displays, beautiful or bizarre, include the most varied antics and are usually accompanied by sounds made with the voice, wings, or collapsing air pouches. In a large, compact assembly, novices establish posts on the outskirts and with advancing years and experience work toward the center. The competing males are friendly, mildly aggressive, or sometimes belligerent; if they frequently injured one another, the assembly could not persist. When a female arrives and moves boldly through the company, posturing and calling intensify as the males vie for her attention, but they cannot compel her. She chooses freely usually one of the older, more experienced performers at the center of the group, a bird who has proved his ability to survive amid nature's perils and reveals his vitality by the splendor of his attire and the vigor of his displays. This, varying with the species and the size of the assembly, is the situation to which we owe a large share of avian splendor.

A psychic trait that birds share with humans intensifies the effects of sexual selection; both tend to prefer the bigger, the more numerous, the more colorful or ornate. Birds respond strongly to supernormal stimuli, as when an Oystercatcher tries to incubate an artificial egg too big for her to sit on, or a Ringed Plover chooses eggs more heavily marked than her own. Over the generations, female birds must have preferred some outstanding characters of males—colors, ornamental plumes, or displays—that had mutated to exceed the norm. A frequently chosen male in an assembly sires many more offspring than can a monogamous male attending the nest of a single mate, with the result that genes of such chosen males spread rapidly through a population, until today we behold the most gorgeous plumes, the strangest feathers, the most curious wattles, or the oddest antics. Birds burdened with extravagant finery demonstrate their competence by managing to survive despite their encumbrances. We can imagine a female thinking: "If that handsome bird can keep himself alive and fit with such an impediment, he must be a desirable father for my nestlings."

Males compete with one another to be chosen by females. Intrasexual selection coexists with intersexual selection but in courtship assemblies the latter predominates. Some ornithologists disagree; doubting that birds have aesthetic sensibility, they assert that the primary function of adornments is to advantage males in direct conflicts with other males. It is true that, before human warfare became mechanized slaughter, warriors adorned themselves, often profusely, for battle; but victory owed more to prowess than to plumes. This interpretation of the function of adornments appears to attribute aesthetic sensibility to male birds but not to females, without offering any proof of this improbable difference. Moreover, the fine details of a male's attire, discernible by a female deliberately choosing a partner, would hardly be noticed by males engaged in a fluttery skirmish. The inverted displays of some birds of paradise, including one of the loveliest of all, the Blue or Prince Rudolf's, convincingly refute the view that lavish ornamentation is a product of intrasexual selection; a bird trying to intimidate another would hardly hang head-downward before him in a submissive attitude, but he might assume this posture as the most effective way to display his splendor to a female.

The bowerbirds of New Guinea and Australia furnish additional evidence that the adornments of males serve primarily to win females, not to awe other males. These unique birds, without counterparts anywhere in the world, prepare a diversity of stages or bowers to which they attract females. The stagemakers, including the Tooth-billed or Stagemaker Bowerbird and Archbold's Bowerbird decorate with leaves or ferns a cleared area of ground. Others build with carefully interlaced sticks. Avenue builders begin by covering a small plot of ground with a thick mat of sticks crisscrossing in all directions. Into this platform they insert upright twigs in two parallel rows, a few inches apart, walling a long, narrow space, an "avenue" open at both ends. On the platform outside the avenue they collect diverse objects: fruits, shells, bleached bones of small animals, fragments of colored glass or pottery, and other artifacts, their choice of adornments varying with the species and the locality. Among the maypole builders is the Golden Bowerbird of Queensland, who arranges sticks around two upright saplings growing close together, with a horizontal branch or vine bridging the space between them to form the tallest of the bowers, which he adorns with flowers, ferns, lichens, fruits, and seeds. The hut builders and garden makers arrange sticks or stems around a sapling to make a dome-shaped or conical hut open on one side, facing a garden adorned with flowers, fruits, or other small objects. The flowers and fruits are replaced when they wither or decay.

The most talented of all bowerbirds is the most severely plain member of the family, the Brown or Vogelkop Gardener, who inhabits the Vogelkop Peninsula at the western end of New Guinea. The sexes of this ten-inch (25 cm), unadorned, olive-brown bird are alike. We can imagine the astonished delight of the naturalist-explorer Odoardo Beccari when he first laid eyes on one of this bird's bowers in the dark undergrowth of the Arfak Mountains in September, 1872. The pavilion, a fit abode for Oberon and Titania, was so carefully described by its discoverer that we can do no better than to read his account:

The Amblyornis selects a flat even place around the trunk of a small tree that is as thick and as high as a walking-stick of middle size. It begins by constructing at the base of the tree a kind of cone, chiefly of moss, the size of a man's hand. The trunk of the tree becomes the central pillar; another

Blue Bird of Paradise (male in inverted display)

whole building is supported by it. On the top of the central pillar twigs are then methodically placed in a radiating manner, resting on the ground, leaving an aperture for the entrance. Thus is obtained a conical and very regular hut. When the work is complete many other branches are placed transversely in various ways, to make the whole quite firm and impermeable. The whole is nearly three feet in diameter. All of the stems used by the Amblyornis are the thin stems of an orchid *(Dendrobium)*, an epiphyte forming large tufts on the mossy branches of great trees, easily bent like straw, and generally about twenty inches long. The stalks had the leaves, which are small and straight, still fresh and living on them—which leads me to conclude that this plant was selected by the bird to prevent rotting and mould in the building, since it keeps alive for a long time. . . . Before the cottage there is a meadow of moss. This is brought to the spot and kept free from grass, stones, or anything which would offend the eye. On this green flowers and fruits of pretty colour are placed so as to form an elegant little garden. The greater part of the decoration is collected round the entrance to the nest.

Thomas Gilliard, from whose book this quotation is taken, added that Beccari found small, applelike fruits, rosy fruits, rose-colored flowers, fungi, and mottled insects on the turf. Years later, Dillon Ripley found other bowers, larger and slightly different in form but all neatly made. In front, the garden, which resembled a carefully tended lawn, was adorned with collections of flowers and fruits segregated as to color. A few simple experiments demonstrated that the Brown Gardener has definite preferences, will not accept every color, and insists that each little heap contain only a single shade. Ripley dropped upon a garden a pinkish begonia, small yellow flowers, and a pretty red orchid, then hid himself to see what the bird would do. Returning, the gardener promptly threw aside the yellow flowers. "After some hesitation and a good many nods and looks and flirts of the tail," the begonia was also cast away. Perplexed by the red orchid blossom, the gardener took it from one pile of fruits or flowers to another, trying to find one that it matched. Finally, with many flourishes, he laid it on top of some pink flowers. The two colors contrasted, but this was the best match that he could find. In addition to flowers, fruits, and fungi, these Brown Gardeners collect pieces of charcoal and black pebbles. Other populations of the same species, on different mountain ranges, have more sombre tastes, preferring black or brown objects to more colorful things. One has only to see a colored photograph of a bower of the type described by Beccari and Ripley to be convinced that the Brown Gardener has a refined aesthetic sense.

In addition to regional diversity in bower construction and decoration in a single species, individual differences are frequent, raising the question of whether bower style is genetically determined or culturally transmitted, like the song dialects of birds and human arts and customs. In support of the second alternative, it is known that, during a long adolescence, bowerbirds of several species spend much time watching adults at their bowers, that their first constructions are rudimentary, and that only by much practice do they become proficient in building and decorating their bowers. These observations are highly pertinent to the wider problems of learning, cultural transmission, and general intelligence in birds. What do tool using, distraction displays, patterns of cooperative breeding, and other aspects of avian behavior owe to learning and cultural transmission, and how has this affected the evolution of innate behaviors?

Bowerbirds are of critical importance for the question of whether intrasexual or intersexual selection has played the major role in the evolution of males' adornments, whether these serve primarily to attract females or to daunt rival males. Dawning intelligence is too often destructive, in birds as well as in primates great and small. Male bowerbirds try to diminish neighbors' chances of attracting females by wrecking their constructions and stealing their display things while the owners are temporarily absent. But bowerbirds cannot pick up their bulky structures to flaunt in the faces of competitors as a display of superiority. When a female visits a bower, the owner often holds a fruit or other object from his collection in his bill while he displays to her. Satin Bowerbirds with bowers that are well made and most profusely decorated, especially those with most snail shells, blue feathers, and yellow leaves, attract most females. Likewise, the mimetic skill of older males strongly appeals to females, who appear to be won by performances that are both vocally and visually superior. The males' personal appearance can hardly fail to influence the females' choices, for some are handsome birds.

Thomas Gilliard called attention to the fact that, among bowerbirds, the plainest males have the most ornate pavilions, as is clear when we compare the exquisite garden of the unadorned male Brown Gardener with the random collection of miscellaneous oddments on the platforms of some of the more handsome species, such as the Satin Bowerbird, whose silky black plumage glistens with iridescent tints of purple, violet, and blue. The adornments that help to attract females have been transferred from the actor himself to his stage setting. When we review all the evidence, we can hardly doubt that intersexual selection—female choice—is responsible for the elegant plumage and striking displays of males, and the argument is clinched by the bowerbirds, who build and adorn structures that attract the opposite sex and who make the most charming bowers to compensate for their own plainness. Moreover, I suspect that they enjoy building and maintaining their elaborate constructions, possibly even taking something like pride in the appearance of their bowers.

After visiting one or more males to inseminate their eggs, female bowerbirds, like female grouse, hummingbirds, manakins, many cotingas, and most birds of paradise, go off alone to build their nests, incubate their eggs, and rear their young without male assistance.

Beautiful plumage is certainly not restricted to species in which males take no interest in nests. Many birds that live in pairs for a season or much longer are lovely. Probably sexual selection, unilateral or mutual, promotes their colorful attire but to what degree is less clear than in the case of birds that do not pair. Males' participation in nest attendance restrains extravagances that might reduce their efficiency. The male Resplendent Quetzal, widely admired as the New World's most elegant bird, incubates and feeds nestlings despite his greatly elongated upper tail coverts, which are frayed or broken by bending and friction on innumerable passages in and out of his nest in a cavity in a decaying trunk. The very diverse color patterns of the many small, beautiful, closely related tanagers that flock together in tropical American rain forests help to preserve specific distinctness by preventing mismating. Sexual selection, which tends to increase the beauty of birds, and natural selection, which favors streamlined efficiency and cryptic coloration, often pull in opposite directions. The appearance of birds is often a compromise between diverse factors.

Birds, nature's chief musicians, produce by far the greater part of its audible beauty, cheering forests, meadows, and gardens with their melodies. The adjectives we use to describe their songs suggest their rich diversity. "Joyous," "ebullient," "ecstatic," "martial," "plaintive," "melancholy" convey the impressions they make upon a sensitive hearer, without implying that they denote the feelings of the feathered singers. Their songs proclaim the possession of territory and attract mates, a fact that explains their prevalence and diversity but not their beauty; distinctive calls, no matter how harsh or grating to our ears, might serve the same ends in creatures devoid of aesthetic appreciation. Our delight in the songs of birds suggests that they affect the singers as they affect us. Without mirrors, birds cannot admire their own beauty, and it might not occur to them that they look much like others of their species and sex, but they can hear their own voices.

Singing, not always territorial, may be social or solitary. Many birds sing all together, in a flock or communal roost, seeming to delight in hearing themselves and their neighbors. A versatile singer inventing a new tune repeats it over and over, and others may copy it. Jays, who lack loud songs, sometimes rest alone in a tree and continue for minutes on end to sing pleasantly in an undertone. Such *sotto voce* medleys appear to lack social or biologic significance; the jay sings for its own comfort or enjoyment; as a human hums a tune when alone and falls silent when another person appears. I venture to assert that if birds find no pleasure in singing, they are incapable of enjoyment, and if they find no joys or satisfactions in their lives, all their efforts to survive and reproduce are barren. They certainly do not sing to delight us, who appeared on Earth ages after they did.

Scattered worldwide among the families of birds are a number of mimics. Their medleys are rarely as beautiful and soul-stirring as the pure strains of many birds who sing more coherently. Often we admire the range and fluency of the mimic's voice more than the bird's musical taste; such a bird entertains, and challenges us to test our bird lore by identifying the originals being copied. Vocal mimicry is important because of the light it throws upon birds' minds. It reveals that birds take an alert interest in the sounds they hear, including many that appear to be unrelated to their basic vital needs. Moreover, it reveals that their behavior is not al-

Horned Lark

ways strictly controlled by their genes; by choosing to imitate this sound or that, by varying the sequence of their own or borrowed notes, they demonstrate that they enjoy a measure of freedom, which they also exercise when they choose partners in reproduction.

Superior song requires an elaborate vocal organ, the syrinx, which did not evolve simply because birds may enjoy hearing themselves sing. For the evolution of its complex musculature, an adequate selective agent was indispensable, and I can think of none except sexual selection, which appears to be as adequate to promote melodious song as to promote beautiful plumage. Both depend upon favorable genetic mutations and their increase in a population by preferential mating. Just as preference for pleasing sights or sounds differs among individual humans, so it may be with birds. Birds of paradise, with magnificent plumage but voices that human hearers do not applaud, are visually oriented. The undistinguished plumage and mellifluous songs of some birds, including a number of thrushes, probably reveal the females' preferences in unilateral selection. When voice and appearance weigh equally with females, the males should be both colorful and songful, as in orioles. Mutual selection tends to make the sexes similar in voice as in plumage, as in wrens. Often the partners join in duets.

Evidence that sexual selection promotes melodious song as well as gorgeous plumage, and by inference the organs that produce superior song, continues to accumulate as more birds are carefully studied in the field. A

Mandarin Ducks (female, left; male, right)

good example is Clive Catchpole's work with the Sedge Warbler. This plainly attired European bird's songs, often up to a minute long, are composed of a repertoire of fifty syllable types, so arranged that no two compositions are alike. Catchpole regarded these recitals as "the acoustical equivalent of the Peacock's train, an extravaganza whose only possible function could be to influence female choice." Warblers with the most complex songs win females earlier than do their rivals with smaller repertoires. After acquiring a partner, they cease to sing, breaking their silence only if they lose their mates. They defend their territories by visual threat displays and active aggression. Moreover, they find much of the food for themselves and their families beyond their territories, all of which supports the conclusion that the Sedge Warbler's exceptionally elaborate singing is a product of intersexual selection.

In addition to the high aesthetic value of bird song, it might be said to have moral value. Birds often settle their disputes by voice instead of by fighting, as by countersinging. Birds in adjoining territories sing alternately back and forth with similar verses and, recognizing that they are in no danger of aggression by well-known neighbors, they desist from attacking one another. In addition to their beauty and melody, the rarity of vicious fighting among songbirds, especially those constantly mated in mild climates, makes them attractive to thoughtful watchers.

Birds prefer sexual partners, momentary or permanent, with attractive plumage and/or appealing songs. They create beauty, indirectly, by their choices of fathers for their nestlings, and directly, by singing melodiously or decorating their bowers, as by bowerbirds. These criteria are similar to those by which we judge the aesthetic sense of humans. What further evidence that birds have an aesthetic sense could we demand, short of direct experience of their feelings, which we can have neither with birds nor with people? Aesthetic sensibility probably varies enormously among the nine thousand species of birds. As among humans, some are more sensitive to visual beauty, some to auditory beauty, and some may be aesthetically insensitive. But denial of aesthetic feeling in birds as a whole, and of their ability to choose between degrees of beauty, leaves us with no satisfactory explanation of ornamental plumes that would be vetoed by natural selection, nor of extravagantly profuse singing. The capacity to have life quickened by colors, forms, sounds, or movements, most strongly when they are most beautiful, harmonious, or rhythmic, appears to be the neurological foundation of the aesthetic sense.

References: Catchpole 1980; Darwin 1871; Gilliard 1969; Ripley 1940; Skutch 1992 (this book contains a full bibliography).

Chapter 9

❦

Dissimulation

ANIMALS PROTECT their progeny by diverse means. Ungulates whose young are highly mobile soon after birth race away from danger with them. Some social bees and wasps attack an intruder in angry swarms, inflicting painful stings, caring not how many of their multitudinous individuals they lose to save their hives or vespiaries with precious broods. Mammalian dams with suckling young, especially the big carnivores, are notoriously dangerous to approach. With sharp talons, large raptors may severely lacerate someone who climbs to their eyries. Even birdlings no larger than a sparrow may nip fingers placed upon their nests. More frequently, instead of attacking potential enemies that might kill them, parent birds try to lead unwanted visitors away. Of all animals, only birds are known to have the finesse to lure dangerous intruders from eggs or young by a display that frequently saves their progeny without great risk to themselves. They have the advantage of being able to escape, by timely flight, the flightless predator that they tempt to pursue them.

Probably most people who wander much through woods and fields have occasionally witnessed this stratagem. Suddenly a bird bursts from a point on or near the ground and flutters or limps wildly over it, beating its wings in an uncoordinated way, appearing unable to fly or even to walk or hop effectively, making you fear that you have inadvertently trodden upon it. If you follow, it will struggle ahead for a few or many yards, then suddenly "recover" and fly away, perhaps with a sharp *chip*. If you retrace its course and search carefully, you may find a nest with eggs or young.

In all parts of the world, a wide diversity of birds from little songbirds to great Ostriches practice such displays. Often they are called "injury feigning," but students of bird behavior object to this designation because it implies that the actors are consciously playing a part or dissembling, which may or may not be true. A noncommittal term for this behavior is "injury simulation," and it belongs to the category called "diversionary displays" or "distraction displays" because it diverts the would-be predator from the nest or young to the escaping parent or distracts it from the primary object of its search.

The psychology of injury simulation has been variously interpreted. Too readily deceived by the appearance of a very realistic act, observers, including Herbert Friedmann and Edward A. Armstrong, have imagined that the distressed parent is in agony or having a fit. They have believed that the conflict between fear, impelling the bird to flee, and parental devotion, binding it to eggs or young, resulted in muscular inhibition and uncoordinated movements. The displaying bird's behavior was "a compromise between fear and reproductine emotion"; it was "in pain," "deliriously excited," or "hysterical"; its behavior patterns were more or less disorganized.

After having watched injury simulation in a considerable range of birds and read about it in many others, I began to doubt this conventional explanation of the phenomenon. I could see no conflict between self-preservation and devotion to progeny when, by the skillful performance of injury simulation, a bird can save both its brood and itself, whereas, if it sticks stubbornly to its nest, it will probably lose both its offspring and its life. I came to view even the wildest acts of injury feigning not as disorganized behavior resulting from emotional conflicts but as a complex, innate behavior pattern that required perfect muscular control and cool judgment to avoid being caught by a pursuing predator, while remaining close enough to lure it onward with the prospect of a meal. Scarcely anything a bird does needs such prompt calculation in a complex, potentially hazardous situation.

My conviction that injury simulation is not a consequence of a crippling emotional conflict but an innate behavior is based upon a number of facts. The first is that distraction displays, including injury feigning, are not determined by the strength of parental attachment but by the type of nest and its situation. Typical injury feigning is most frequent in birds with open nests, or closed structures with a side entrance on or near the ground, although wood warblers that nest high in trees may perform along a lofty bough or drop to the ground to grovel over it. Birds that nest in holes in trees, termitaries, or burrows in the ground rarely give distraction displays. A snake or other predator that creeps into a burrow traps a parent it finds within. If the adult manages to slip past the marauder and escape, or if it arrives to find an enemy intruding, any display that it might make would not be seen by an animal with its head inside the tunnel.

Moreover, a bird does not feign injury unless it finds a clear stage for its act; it avoids entangling itself in dense vegetation. Seabirds that nest in crowded colonies could not flap over the ground in apparent helplessness without stirring up a disastrous uproar. Kingfishers and other birds that nest in or on river banks would find the water an unsatisfactory stage for injury simulation. Although Buff-rumped Warblers belong to a family in which injury feigning is frequent and realistic, their nests often face streams; I never saw one feign until I found my thirty-fourth nest, beside a rock in a pasture that offered an excellent stage, on which the female gave a number of displays of moderate intensity, creeping over the close-cropped grass with spread, fluttering wings. We have no reason to suppose that birds who never feign are less strongly attached to their nests than are those who do. Parent kingfishers and motmots have clung to newly hatched young while I uncovered their subterranean chambers and lifted them out. When released, they might protest loudly but they did not display.

Proof that injury-feigning birds are in full control of their faculties is provided by those which alternate normal flight with simulated injury. The Common Pauraque is a nightjar or goatsucker widespread in tropical America. It incubates its two eggs or broods its chicks on leaf-strewn ground beneath a thicket, with no indication of nest building. As a person approaches the brownish sitting bird, it rises lightly in the air to descend a few yards away and grovel as though injured among brown dead leaves, pressing its breast to the ground, beating the earth with its wings, while uttering low, soft, croaking sounds. If you follow, it bounces easily to a point farther onward and repeats the act, and so on, until it has led you to a satisfactory distance from its progeny.

Wishing to see what was in a White-tipped Doves' nest, well above my head in a coffee plantation, I first raised the long stick to which I intended to attach a small mirror. Reluctant to abandon whatever it was covering, the brave bird raised its wings straight above its back in an attitude of defiance, revealing the beautiful cinnamon of the underwing coverts, then struck the stick's end with such a resounding whack that I feared it had broken a bone. When I persisted in my attempt to see the nest's contents in the mirror, the dove dropped to the ground and fluttered over it so haltingly that, following, I suspected that its wing was indeed fractured. For two hundred feet (60 m) the bird led me over the cleared ground to a thicket at the plantation's edge, over which it flew without difficulty, to alight on a log in an open space and resume its "broken-wing" ruse, trying to lure me still farther from its nest. Unable to transcend the barrier as readily as the dove did, I returned to its nest, where my mirror revealed a half-grown nestling.

Other birds behave differently when they reach a barrier. A male Plain Antvireo, driven from his nest in a sapling in the rain forest, dropped from newly hatched nestlings to the ground. Here he turned repeatedly to remain within the small clear space amid the undergrowth, where almost at my feet he crept, beating against leafy ground dark wings that revealed

Common Pauraque (female)

next to the body white bands that I had not previously noticed. He held my attention downward, away from his nest above my head.

Parent birds adapt their strategy to the nature of the animal that threatens their nest. P. A. Taverner described how a Killdeer nesting on open prairie in Manitoba reacted to an approaching horse or cow, who might accidentally crush its eggs. The bird lay low until the grazing quadruped was almost over the nest, then flew suddenly into its face with a great outcry, making the animal stagger back and circle the spot in confusion. After "a few dives and expostulations," the Killdeer returned quietly to its eggs, its objective accomplished. When a man or dog approached, it flew to meet him from afar and entertained him with displays made more realistic by appearing to lie helplessly on its back, although actually it rested more safely on its belly. When it had lured the intruder to a satisfactory distance, the Killdeer returned to its nest.

Injury feigning is not restricted to birds driven from nests where they have been sitting, incubating eggs or brooding nestlings. One of the most dramatic episodes that I have seen occurred one sultry afternoon while I loitered beneath palm trees in a seaside park in Costa Rica. While I watched

Killdeer

a family of Black-striped Sparrows consisting of parents, two juveniles with spotted breasts, and a fledgling of a later brood barely able to fly, a nurse girl came along with a child and a dog. Spying the birds on bare ground beneath shrubbery, the dog rushed toward them. The fledgling, far from cover, appeared to be on the point of falling into the beast's jaws, when in the blinking of an eye a parent intervened. Fluttering over the ground just ahead of the menacing fangs, it led the dog swiftly away from its helpless offspring. The dog was still eagerly following the parent sparrow when they passed from view amid the shrubbery.

As we would expect, most recorded acts of injury simulation have been performed in front of people and the dogs that accompany them. A canine can rarely resist the temptation to pursue a seemingly injured bird in hope of a meal that, as far as I know, always eludes it. Once, at dawn, I watched an injury-feigning Common Pauraque toll a lumbering opossum in a wide semicircle around the point on the ground where his mate covered two eggs. Except dogs and myself, the only other mammal that I have seen elicit injury feigning was an agouti, to whom a Chestnut-backed Antbird displayed in tropical rain forest. But, more interested in eating seeds that littered the ground around the antbirds' nest than in the parent bird's act, the vegetarian rodent was not enticed to follow him. Otters, weasels, stoats, foxes, coyotes, and deer have also been reported to incite injury-feigning displays. Although snakes are major predators on eggs and nestlings, parent birds are more inclined to attack them than to lure them away. Rarely, the display is directed to a larger, nonraptorial bird, as when a Kentish Plover, watched by Edmund Selous, attempted to lure an Oystercatcher away from its nest. With no desire to catch the plover or eat its eggs, the bigger bird did not follow far.

One might suppose that injury simulation in front of an active predator is a perilous ruse; but it is almost confined to displays before mammals, from which an alert bird can easily escape by flying, except from humans who could kill it with stones or other missiles. In a survey of risks involved in distraction behavior and mobbing, Tex Sordahl found that mobbing—when birds gather around and harass a potential enemy, such as a drowsy owl in the daytime, with cries and feints of attack—is much riskier than injury feigning. He found only three published accounts of fatalities while feigning. Dianne Brunton saw a fox approach from behind and catch a calmly incubating Killdeer who, taken by surprise, started to display before it had reached a safe distance. Apparently, a feigning bird is in greater danger of being seized by a second bird while its attention is centered upon the object of its display than by the latter; as when a Mourning Dove, performing before Russell Balda who had approached its nest, was struck and killed by a Loggerhead Shrike. Despite such occasional mishaps, the wide diffusion and persistence of injury simulation is proof that it saves many eggs and young with little danger to their parents, whose loss nearly always dooms dependent progeny. If this were not true, natural selection would have restricted or eliminated this practice.

Another diversionary display that involves dissimulation is the "rodent run," or "rat trick." Alighting in front of an intruder, a parent bird runs ahead of it with foreparts depressed in a hunched posture. With its tail depressed, the bird's trailing, quivering wings simulate the rapidly moving hind legs of a small rodent. Often, too, the bird's fluffed-out feathers resemble fur, and it may even squeal like a mouse. Looking and acting as much like a small rodent as a bird can, it scurries away in a zigzag course, at intervals pausing briefly to look behind and see whether the enemy is in pursuit. When followed, the bird may continue its rodentlike running for hundreds of feet over the open tundra. The ruse is most frequent among sandpipers, plovers, and other shorebirds that nest far in the north, where rodents, principally lemmings, are a main food of predators. It may have evolved to reduce predation on nests by the Arctic Fox, which probably is readily tempted to pursue a bird that simulates its preferred prey, and so is lured away from the performer's eggs or chicks.

At lower latitudes variations of the rodent run have been observed in wren-warblers of Australia and Green-tailed Towhees of the western United States. When one of these towhees scuttles away from its nest in a fast,

Great Horned Owl

even motion with tail elevated, it resembles one of the chipmunks abundant in the sagebrush where the towhees' nests are situated, and it may deflect coyotes from them. Among tropical birds, Black-striped Sparrows give a somewhat similar performance. Although ordinarily they hop over the ground with feet together, when displaying they run with alternately advancing feet, body depressed in a hunched attitude, tail held low, and wings more or less spread. I have seen a brown ground snake and a rat or large mouse pursue a Black-striped Sparrow escaping in this fashion. Black-striped Sparrows and Black-headed Brush-Finches run in much the same way ahead of fledglings who have incautiously ventured onto open ground, leading them to cover. It may seem strange that a bird should lure away predators and guide its fledglings to safety with similar movements, but the two occasions have a common feature—progeny in peril.

A bird may feign injury or give some other distraction display at any stage of its nesting, but it is most likely to do so around the time its eggs hatch and when its young leave the nest. A curious aspect of these performances is their unpredictability, in species and in individuals of the same species. Other activities of nesting birds, from building through incubation, nurture of the young, and the departure of the nestlings, run a regular course; you can count upon their occurrence at appropriate times. But injury feigning appears to be capricious. In the highlands of Costa Rica, I studied a number of nests of the pretty Slate-throated Redstart, a wood warbler, all strung out in niches in a bank above a path that ran between a pasture above and montane rain forest below. All the roofed structures with side entrances faced the path, which offered all these birds a good stage for the feigning act. Yet of seven whose nests survived at least until the eggs hatched, five performed for me and two never did.

Although I have watched injury simulation in a diversity of birds, and many more such displays have been reported by others, nearly always the observations were incidental rather than planned; usually the displaying bird surprised the observer. The only carefully controlled experimental study of this behavior of which I am aware was conducted by Carolyn A. Ristau with Wilson's Plovers and Piping Plovers. These shorebirds nest in slight depressions, scraped into the sand and often unlined, on open beaches or dunes, where they are easily watched. The experimental procedure involved an observer inconspicuous at a distance and an "intruder" who walked past a nest where a plover incubated. If he or she revealed interest in the nest, as by stopping to look closely at it, this person was considered "dangerous." If he or she simply walked by at a distance, without gazing or otherwise revealing awareness of the sitting bird, he or she was "safe." The "dangerous" intruder caused the plover to leave the nest with a broken-wing display that was obviously intended to lure this person farther from the nest. Most parent plovers soon learned to recognize, by distinctive clothing or otherwise, humans who had been "safe" on previous appearances and remained on the nest, perhaps stretching up to look around, when they walked past. For details of these carefully analyzed observations and a discussion of their implications, I refer the reader to Ristau's paper listed in the bibliography. With the usual caution of a scientist discussing a controversial subject, she concluded: "These experiments are only a beginning in the exploration of whether and to what extent plovers are intentional creatures [aware of what they are trying to accomplish]. The results so far suggest that they are."

Although the forms of distraction displays appear to be inately determined, they are not so strictly programmed by the genes as are other activities connected with reproduction; their occurrence depends more upon the minds or fluctuating feelings of individual birds. They appear free to perform or not to perform, as they please. The variance among individuals in the occurrence or frequency of distraction displays may give us a clue to their origin.

Brains, the minds they support, and the sensory organs that report to them evolved to adjust animals to the variable external world. Developed minds sometimes devise ways to improve this adjustment and increase survival and reproduction, or to make living more enjoyable, as by playing. It should be highly advantageous to a species to perpetuate improvements in descendants and diffuse them through the species. Two methods of achieving these results are available: they may be passed directly to the progeny and associates of the inventor by teaching or example, becoming traditional; or they may be encoded in the genotype (an organism's complement of genes), thereby becoming more firmly established in a species. The first method is universal in our own species, the foundation of our cultures, and may be more or less prevalent in other social animals; the second is more problematic.

The genes that control an organism's growth, form, and physiology, and to a greater or lesser degree its behavior, are embedded in its tissues and have little direct contact with the world around it. They may be reached by hard radiations or heat, which cause mutations that are random in the sense that they do not occur in response to the organism's needs. They are harmful more often than helpful; only a minority contribute to the adaptive evolution of a species. In contrast to the genes, the mind, through its senses, is in close contact with the environment. One would expect that it would be advantageous to a species to have any improvement of mental origin firmly encoded in the genotype, that it might be transmitted directly to the individual's descendants, along with all its other characters; that evolution would have developed a means for such direct transference. Strangely, this has not happened; such transmittal would entail the inheritance of characters acquired during the lives of individuals—Lamarckian evolution, for which we lack sufficient evidence. Although acquired characters may not directly become hereditary, they may nevertheless do so indirectly, by the method called organic selection, or, more explicitly, genetic assimilation, because by a roundabout course a behavior originated by individuals becomes encoded in the genotype.

Like play, injury simulation and other distraction displays of birds involve no special organs or modifications of organs, and probably no special nervous connections. Every element in the situation—nest, approaching predator, terrain, the bird itself—is immediately apparent to the performer. Activities of this sort appear much more likely to originate in the mind that guides them than in genes that have no direct contact with the external circumstances. We do not know when injury feigning began; the fossil record offers no clue. It is most unlikely to have started among ancestral reptiles, unable to escape the enemy by flight and apparently lacking the intelligence to devise such a ruse; but its wide diffusion among living birds suggests that it originated ages ago.

I suggest that the first step in the development of injury feigning was taken when some unknown bird, strongly attached to its eggs and young, abandoned them with slow reluctance, probably quivering its wings, as a predator approached its nest. Attracted by the prospect of catching the slowly retreating parent, the enemy followed it away from the nest. The bird remembered this, and when confronted by a similar threat, departed its nest in the same manner, possibly intensifying the movements of its wings, making itself more conspicuous and attractive to its pursuer. The progeny saved by these displays might behave like their parent. In successive generations the display might improve, more closely simulating the wild flutterings of a crippled bird, becoming more tempting to a carnivore. Even a slight advantage in the success of nests and survival of fledglings should make the lineage that practiced this ruse increase more rapidly than others that lacked it. The displaying birds would thrive and multiply. In the course of many generations, some of the randomly occurring genetic mutations, which can alter any aspect of an organism's structure or function, would support this behavior, perhaps improve it, until it became firmly established in the genotype. Genetic assimilation would have made heritable a display that originated in a bird's mind. Because birds are now so well equipped with hereditary patterns of behavior, they have less need to devise new solutions to their problems.

The wide diffusion of distraction displays and their differences in detail suggest that they have originated independently in different families or genera, as the nesting habits of these birds made them serviceable. Changed circumstances may cause their decay, but their stability is attested by the Galápagos Dove, which like many other pigeons readily feigns injury, although it has long inhabited certain islands of the archipelago where its nests were not subject to predation before the recent arrival of humans and the quadrupeds they brought with them.

Although I may have exaggerated the role of mind in the origination of injury feigning and some other distraction displays, I have no doubt that their successful performance before animals of various kinds in different settings requires more mental alertness than almost anything else a bird does. To preserve its life while dissimulating before a ravenous predator, a bird needs sharp wits.

Although the most widespread, injury feigning is not the only kind of dissimulation practiced by birds. In Venezuela, a pair of Striped-backed Wrens tried to save their eggs from a collector by a very different deception. While George Cherrie was cutting the branch that supported their bulky nest with its side entrance, they neither scolded nor threatened him but tried to mislead him by busily carrying material into a deserted nest nearby, as though they were preparing it for a brood. When this ruse failed to divert Cherrie's attention from their occupied nest, the wrens returned to it and packed the doorway with down from the silk cotton tree and with soft feathers, so rapidly that before the nest was on the ground, no entrance was visible. For a moment, the collector failed to find the eggs.

More successful deception was practiced by a female African Marsh Harrier studied by Robert Simmons. After the loss of her mate left three feathered nestlings without a father, two males fought for his territory.

The younger won, and after initial hostility began to feed the young harriers abundantly, while ignoring their mother's pleas for the prey he carried. Thereupon, she deserted her territory to solicit and mate with the defeated rival of the foster father. Despite the lateness of the season, her new consort started to build a nest that she never occupied. When he gave her food, she ate part of it and carried the rest to her young, who had now fledged. The deceitful female enjoyed the rare situation of having two unrelated males providing food for her progeny, while she directly provided nothing.

Deception was also practiced with success by a Cackling Canada Goose at a lake in Wisconsin, where Philip Whitford studied Giant Canada Geese. One day a family of two adult and six juvenile Todd's Canada Geese (another race of this widespread species, which varies greatly in size) flew down to the lake and took possession of the feeding site, keeping all the other geese at a distance. A goose family is usually closed against strangers, but the wily Cackling Canada Goose managed to intrude temporarily into it. After being displaced from the food by the advancing line of newcomers, this goose ran around behind and entered the line, neck pumping and calling. When, upon reaching the food pile, the family joined in a triumph ceremony, waving their heads and necks together and calling, the intruder participated in this ritual usually reserved for a family, performing just as they did. By clever dissimulation, the Cackling Canada Goose managed to eat with the Todd's Canada Geese for half an hour before the family gander detected the deception and chased the interloper away.

Dissimulation—appearing to be what one is not—is a means widespread in the animal kingdom of escaping predators, and even of deceiving victims of predation. Familiar examples, falling into the category of procryptic, or concealing, coloration, are the mottled brown nightjar that appears to be part of the brown leaf litter where it rests, the green parrot that resembles the green foliage of the tree where it forages, the gray moth that looks like the gray bark to which it clings. A widespread mode of dissimulation among butterflies is Batesian mimicry—a palatable butterfly escapes predation by closely resembling another species whose bad taste deters predators. In all these cases and innumerable others, the dissimulation is innate and passive; the creature cannot change its colors to suit its circumstances. More flexible is the dissimulation of birds such as ptarmigans, who by molting become white to resemble snow in winter and brown to blend into the ground cover in summer. These alterations of attire are also genetically determined. Still more flexible and responsive to the immediate situation are the color changes, effected by the expansion or contraction of chromatophores in the skin, of certain lizards, frogs, fishes, and mollusks to match the surface on which they rest. These active changes appear to be automatic and involuntary.

More clearly voluntary is the active dissimulation of birds, as in injury feigning, when a perfectly sound parent bird decides to appear crippled, in circumstances that appear favorable for saving itself and its progeny by this deception. Now that field naturalists are becoming aware of the survival value of dissimulation by the active minds of birds, more subtle instances of deceit are being investigated.

References: Armstrong 1942; Balda 1965; Brunton 1986; Cherrie 1916; Friedmann 1934; Ristau 1991; Selours 1927; Simmons 1992; Skutch 1954-55, 1976; Sordahl 1990; Taverner 1936; Whitford 1990.

Chapter 10

Mental Conflicts

A LITTLE, brownish, bare-cheeked Masked Tityra chose for her nest site an abandoned woodpeckers' hole near the top of a tall dead tree, standing in a recent clearing in the rain forest. Only a few feet below her cavity, in a hole carved by a larger woodpecker, four Fiery-billed Aracaris slept. Much bigger than the tityra, these middle-sized toucans with huge, flame-colored beaks were not above eating eggs and nestlings. The tityra strongly disapproved of the presence of such dangerous neighbors. Each evening, as they flew to their high bedroom, she pursued and darted at them, sometimes with her white mate; but they ignored the harmless tityras. By day, when the aracaris were absent, the female tityra covered the bottom of her cavity with a loose litter of small, dying or dead leaves and leaf fragments, on which she would lay her eggs. Finally, as daylight faded, her attentive partner would fly off alone to roost in the forest, leaving her in the clearing.

Left to herself, the female tityra would rest upon the lofty summit of the trunk that contained her nest and the aracaris' dormitory, or upon a nearby eminence. After a while, she would fly toward her doorway, but as she came close to it, and to the doorway of the cavity in which her dreaded enemies were already ensconced, she would lose courage and return to her high perch. Here she stood motionless, a small, solitary figure silhouetted darkly against the fading light above the western ridge, while all the bird world sank into hushed repose. I could imagine the turmoil in her mind while she rested, outwardly so still, beneath the great, open, cloud-veiled sky. She felt impelled to pass the night in her nest, which probably already contained two mottled gray eggs, but she feared to sleep so near the aracaris, even if several feet of solid wood did separate the two cavities. Finally, after even the late-retiring motmots and woodcreepers had sought their sleeping places, and bats were fluttering through the dusk, the lonesome

Fiery-billed Aracari (left) and Masked Tityra (female)

little bird made another move toward the round entrance of the hole, now scarcely visible in the shadowy trunk. But in front of it her courage failed; she hovered on fluttering wings, then rose to rest again atop the trunk. Five or six times she started toward her nest only to turn aside. At last, she winged away to the forest, whither she and her partner had gone together on previous evenings, and vanished amid the dusky foliage. A few evenings later, her eggs' need of warming drew her to them strongly enough to overcome her dread of the toucans.

We are never more acutely conscious than when conflicting impulses contend for control of our behavior. As I watched the tityra resting motionless against the darkening sky, I could hardly doubt that she was conscious of her situation.

On the stony shore of a mountain torrent, I found my first nest of the Black Phoebe, a familiar figure along swift inland waterways from Oregon to Bolivia. The thick-walled cup of gray mud, strengthened by bits of vegetation, was plastered to the vertical side of a huge, overhanging rock, where it was inaccessible to all wingless predators but not beyond reach of rising water. I decided to watch the phoebe incubate her two white eggs,

Masked Tityra (female)

which lay on a bed of soft grasses and feathers. Ordinarily I would have concealed myself, but the stony margin of the wide stream offered neither screening shrubbery nor a level spot to set a blind.

When the phoebe flew down the river and saw me sitting only twelve yards from her nest, she stood on a rock below it, hesitating. After a while she flew up as though to enter it, but when a few inches away, she faltered and dropped back to the low rocks. Several times she started toward her nest without reaching it. While the black bird delayed in this vacillating state, her mate arrived and hovered in front of the nest. His presence gave her confidence; without further delay, she darted up and settled on her eggs in front of him. Then he dropped down to stand quietly below her while she steadily incubated. After his departure, she sat for a few minutes longer, then flew away to catch insects.

When, after a brief absence, she returned, she still lacked courage to go to her nest in my presence, although I had not moved and she had already sat for thirteen minutes in front of me. After many false starts toward her earthen cup, she flew across the river, to return in a few minutes and repeat her wavering behavior. Still fearing to enter her nest, she flew upstream to seek her partner, who returned with her and alighted on the rim, facing inward. Now, without hesitation, she returned to her eggs. For a minute, he stood above her, then dropped down to rest on the rock for a minute or two more.

For the remainder of my five-hour vigil, the female phoebe would never enter her nest unless her mate were close. When he accompanied her return from a recess, she went promptly to her eggs. When she came alone, she stood below the nest with repeated *chips* until he arrived to see her safely on it. Once, while he hawked for insects across the stream, she waited ten whole minutes, meanwhile flycatching from the rocks beneath the nest, or, becoming impatient, crossing the stream to seek him and bring him back with her. When he stood on the nest's rim or fluttered beside it, she promptly entered, but when he stood below it, she made several false starts before she settled on it. He always lingered for a while on the rocks below, for less than a minute to as much as ten, before he disappeared.

Was this behavior, unique in my experience with birds, typical of Black Phoebes or caused by my presence? Did the female refuse to return to her nest in the absence of her consort because she needed him to bolster up her courage? To answer these questions, I returned in the afternoon to watch at a distance of fifty feet but still without vegetation to screen me. Soon the female returned alone, hesitated to go to her nest, then flew upstream, apparently to seek her mate, who did not return with her. A repetition of this sequence had the same result. A third flight upstream was followed by a third solitary return to the nest site, and now, since she could not persuade her usually attentive consort to escort her, she had no recourse but to settle on her neglected eggs in his absence.

Next morning, while during four hours I watched from the more distant rock, the phoebe, becoming reconciled to her watcher, entered her nest alone with little or no hesitation. On only nine of her twelve returns from foraging did the male accompany her. This was apparently about the normal frequency of the pretty custom of seeing her on the nest—a habit practiced, with variations, by many monogamous males who do not incubate.

One aspect of the female phoebe's behavior, her many starts toward her nest, each falling a little short and ending in a retreat, is widespread among birds in similar circumstances; unusual was her insistence on having her mate close by her when she entered. Without having other Black Phoebes' nests to study, I could not learn whether this is an idiosyncracy of the species, but this seems improbable because unnecessary activity at a nest does not increase its probability of success. But certainly the behavior of this female phoebe was very humanlike; the presence of a trusted partner gave her courage to do what she feared to do alone. As in the episode of the Masked Tityra, the stress of opposing drives probably intensified her awareness of her situation, which increased her need of her mate's psychic support.

Unlike some smaller birds, toucans have never struck, pecked, or threatened me when I inspected their nests in cavities in trees. Big Rainbow-billed, or Keel-billed, Toucans, kept from their nest by my visits, relieved their feelings in peculiar ways. One of them, arriving with a white seed for its nestling in the tip of its great, multicolored beak, threw the seed back into its throat with the upward toss of the head that toucans use to swallow their food, then immediately brought the seed up into the tip of its bill again. It repeated this act twenty-three times more, then flew away visibly carrying the seed. On other such occasions, the parent toucan restlessly shifted an insect back and forth between its bill and a foot. More often, when a parent found me and my helper near its nest hole, it perched high in a treetop and continued interminably to complain with sounds like winding a cheap alarm clock.

When I climbed to a Brown Jay's nest in a willow tree beside a lagoon to inspect the nestlings, one of the five attendants darted within a few inches of my head, menacing me. Not daring to strike, it relieved its overwrought feelings by pecking a branch and tearing to shreds leaves of a nearby banana plant. Such redirected aggression springs from the mental conflict between the urge to attack and fear of the consequences.

Although injury feigning is sometimes attributed to emotional conflict, it appears to be in a different category from the foregoing episodes. Whatever its evolutionary origin, feigning is now an innate, adaptive behavior. When a bird has an effective means of saving its progeny, with little danger to itself from the flightless predator that it lures from the nest, conflict of interests is absent. Moreover, the first sign of danger often incites the display too promptly for vacillation, and while acting as though injured in front of a hungry predator, the bird needs to pay strict attention to what it does; indecision at this juncture might be fatal. In the more prolonged mental conflicts described in this chapter, the clash of opposing impulses or drives intensifies consciousness much as friction heats solid bodies. The mental strain finds relief in frivolous activity, as in the toucan and jay, or in vacillating behavior, as in the tityra and the phoebe. In stressful predicaments, we often act in comparable ways.

Avian reproduction demands strenuous effort and faithful performance of duties, rarely equaled by other animals. While embryos develop in a female mammal's body, perhaps unknown by her, she continues her usual activities. While embryos develop in birds' eggs, one or both parent must sit patiently for weeks, warming them, shielding them from strong sunshine and hard downpours, sometimes enduring hunger, restraining impulses to join companions, to fly, or to forage. To nourish her sucklings, a mammalian mother needs only to eat well; processes of which she is un-aware convert part of her food to milk. A parent bird, even when hungry, desists from eating a large part of the food she or he finds, to pass it directly from the bill or mouth to nestlings or fledglings; while feeding young, birds often lose weight. When menaced, a mammalian mother can flee with the embryos in her womb, her sucklings attached to her body as in marsupials, or her mobile young running with her, as in ungulates. A threatened bird cannot escape with eggs or nestlings; it must either aban-don them or risk its life shielding them (not all birds can avoid this di-lemma by distraction displays). In birds, the frequent divergence between self-maintenance and parental duties causes recurrent mental conflicts that other animals are mainly spared. These can have important psychic con-sequences, including the development of something like a conscience, which we might call a protoconscience.

Conscience is a state of mind determined by the conformity of our con-duct to our guiding rules or principles, whether these are socially imposed or self-chosen. When we abide by the rules and discharge our obligations, conscience is more or less tranquil; when our behavior diverges from our principles or we neglect recognized duties, it nags and distresses us, the more acutely the more flagrant our transgressions. To avoid this persistent mental goading, conscientious people undertake distasteful tasks, deny themselves pleasures, even suffer deprivations and hardships. Similarly, birds appear tranquil while they perform closely to their innate patterns of behavior—their guiding rules. When prevented from following these rules, they give signs of distress. To avoid this mental perturbation, while at-tending their families they endure hunger and fatigue, or risk their lives in nest defense that is too often futile. In behavior that resembles that of a conscientious human in trying circumstances, I detect adumbrations of conscience in birds.

References: Skutch 1960, 1969, 1971, 1983a.

Chapter 11

꩜

Intelligent Birds

SUPPOSE THAT, a newcomer to the Galápagos Islands, you watch a Woodpecker Finch pluck a cactus thorn and poke it into a crevice in a tree to chase out a spider that it devours. Never having heard of such behavior, you are excited by your discovery of such an intelligent bird and eager to tell about it. But, as you prolong your visit and see other Woodpecker Finches using probes, and on looking into the matter you learn that this habit of the finch has long been known, you revise your estimate of the bird's intelligence. This appears, after all, to be just one of the finch's innate behaviors, encoded in its genes and revealing no more fresh insight or innovative intelligence than we display when we perform a complex act that we have been taught, such as tying a difficult knot.

We do not know how or when the Woodpecker Finch's tool using began and are not sure how it is perpetuated. Possibly it originated as a fresh insight in an exceptionally bright finch, and became frequent in its equally intelligent progeny, enabling them to exploit a source of food unavailable to surrounding birds, thereby increasing their fitness. If this is correct, the finches' tool using is an example of crystallized intelligence— a habit born of an innovative mind that no longer requires such a mind for its exercise. I surmise that many other facets of avian behavior, such as improvements in nest construction and care of the young, are crystallized intelligence. And we humans certainly live by crystallized intelligence, which is much more common and less highly valued than the fluid or innovative intelligence to which we owe the inventions, the moral and intellectual advances, that are the foundations of civilized life, and with us are transmitted from generation to generation by example and learning.

Because free birds in their ancestral habitats are so well equipped with innate behaviors to meet the contingencies of their lives, they have little need to behave in ways that we can confidently ascribe to fluid intelligence. In interactions with humans, so unpredictable, presenting such novel situations to birds, they are most likely to display intelligence by acting in novel ways—and we are most likely to witness and record such behavior. In many years of watching free birds, I have rarely seen behavior that I was fairly certain was not innate.

The first of these instances occurred while I studied a pair of Gray Catbirds nesting in a barberry hedge at my boyhood home in Maryland. They carefully guarded their nest. Nearly always, when the female left it to forage, her mate came to watch from a nearby branch; when she returned, he retired to a more distant tree to sing. After their nestlings hatched, the parents became boldly protective. When I examined the nest, the male buffeted my head from behind, while his consort struck my hand with her feet or pecked it with her slightly deformed bill. When I held a hand over the nest palm downward, she stood on the back of my hand to shower pecks on it, sometimes drawing a little blood. When I turned the hand palm upward, she would not alight upon it but perched beside it to peck. She appeared to know how the human hand works, and that she was in greater danger of being caught when she stood on the palm than on the back. I doubt that, in the days before bird banding had become as frequent as it now is, Gray Catbirds had had enough encounters with hands to make such knowledge innate. The female with the deformed bill seemed to understand the situation.

In Panama, a yellow-breasted Vermilion-crowned, or Social, Flycatcher started to build a bulky, domed nest with a side entrance in a small lemon tree, where a pair of Scarlet-rumped Tanagers were already feeding nestlings in a substantial open cup. Before she had proceeded far, ants killed the tanager nestlings, leaving their nest vacant, whereupon the flycatcher claimed the empty cup and began to transfer to it the straws and weed stems she had already collected in her nearby site, building up the rim. With grasses gathered at a distance, she roofed the nest, thereby converting an open into a domed structure, in which she laid three eggs. By suddenly changing her plan when her neighbors' nest was left empty and using it as a foundation for her own, she saved herself the labor of hunting for and fetching as much material as the tanagers' nest contained, which was flexible, intelligent behavior. Although I have found at least forty Vermilion-crowned Flycatchers' nests, I have never noticed another instance of such conversion; but at least three of fifty nests of the closely related Gray-capped Flycatcher were finished by roofing open nests of other species.

One morning, as I walked along a trail through the forest in Costa Rica, a Bicolored Antbird of unknown sex followed me, catching insects driven up by my footsteps. Noticing this, I went slowly, deliberately stirring the ground litter with my feet or a stick, and it continued to accompany me. This was the first of many such journeys over an interval of sixteen months, some for a long way through the forest and all, I was confident, with the same individual, whom I called Jimmy, the scientific name of this antbird being *Gymnopithys bicolor*. Sometimes, while I stood watching another bird's nest, this bird would cling low on an upright sapling beside me, calling my attention to itself by soft, questioning notes until I was ready to move onward and provide more good foraging. This antbird would permit me to ruffle its plumage with the short stick I used to stir the litter, but never to touch it with a hand.

Bicolored Antbirds forage mainly with the army ants that swarm over the ground in tropical America, driving from beneath the litter the small creatures that lurk in it and making them readily available to a diversity of feathered ant-followers, who rarely eat the ants themselves. Jimmy was using me as a substitute for the ants, much as anis and Cattle Egrets exploit grazing cattle to stir up grasshoppers for them. Nobody else has, to my knowledge, reported a Bicolored Antbird following a human, not even Edwin Willis, who spent years intensively studying this species in neighboring Panama. Indeed, foraging with terrestrial mammals of any kind appears to be otherwise unknown in all the large antbird family of tropical America. Jimmy was an original genius, intelligently exploiting a profitable method of procuring insects, spiders, and small frogs. In later years in the same part of the woods, other Bicolored Antbirds, possibly Jimmy's descendants, also followed me. Perhaps I witnessed the first steps in the evolution of a new behavior, which might become innate if the shrinking lowland tropical forests where Bicolored Antbirds live are not all destroyed.

In a more domestic setting, a hen exhibited intelligent behavior that was evidently not hereditary. For many years (until a Tayra from the forest carried off the last of them) we had a flock of chickens of the old *criolla* stock, varying in color from white to black with endless intermediate colors and combinations of colors and markings, including some hens and roosters whose elegance left no doubt that they belonged to the aristocratic pheasant family. The personalities of these fowls were as diverse as their plumage, some being peppery and aloof, others mild and friendly.

One of my favorites was a lovely hen called Cercomacra, from her resemblance in coloration to the female of the Dusky Antbird, *Cercomacra tyrannina*. When ready to lay an egg, she would seek me in my study, alight on the table where I wrote or on an arm of the chair where I read, and stay, uttering low, conversational notes until I caressed her, then carried her to a nest box on the other side of the house and draped a sack over the entrance at the front; she preferred to deposit her egg in privacy. Sometimes she would promptly emerge and come to me again, but when

Bicolored Antbird

she began to open her mouth and pant, I knew she would remain in the nest until she emerged, cackling, to announce that she had laid. No other of the hundred hens we raised over the years had this habit, although many liked to have the nest box covered. Probably Cercomacra was among those I had occasionally replaced in a nest when they were restless and would not stay unless they were covered, but I had not taught her to seek me for this service. That was her intelligent way of getting what she wanted, and possibly, also, to show her attachment to the man who fed and protected her.

Len Howard, an exceptionally perceptive observer whose cottage was always open to the birds that frequented her garden, enjoyed unique opportunities for witnessing their intelligent behavior. Newcomers sometimes collided, occasionally disastrously, with the glass in her windows. A Great Tit, Jane, showed much agitation the first time her fledglings approached closed windows. She called her brood to the fanlight through which the birds entered and left, took food in her bill, and showed it to them inside a pane of glass while they perched outside. Trying to reach the offering, they pecked at the glass. Then Jane stuck her head outside to show the food without the transparent barrier, but before they could take it she withdrew behind the pane and held it while they pecked the glass again. After she had done this three times, her young, "with characteristic quickness of Great Tits, understood and were examining the window panes with quaint expressions of interest on their faces." Thereafter, when

Great Tits (adult with young)

they followed their parents into the house, they found their way out through open windows and never banged into closed ones. In the wild, birds learn much from their elders, by example if not by deliberate instruction, but the perils of window glass are not included in their lessons.

One December day, a Blue Tit, new to Howard's cottage, flew in through the fanlight that in cold weather was the birds' only entrance. When ready to depart, the tit fluttered and banged against a window on the opposite side of the room. From outside, another Blue Tit, familiar with the house, noticed the panic-striken bird and, after watching briefly, flew around the building and entered through the fanlight, to which it tried to call the prisoner. When the latter failed to respond, the rescuer flew across the room and touched it, then led it back to the fanlight and freedom—an act of intelligent altruism.

Howard's feathered guests looked to her for protection. On many a morning in the nesting season, she was awakened at five o'clock by a Great Tit flying with loud alarm calls to and fro between her bed and the window, urging her to come quickly and save its young from a marauding Magpie. She would leap up promptly and run outside with a stick to chase away the enemy. After returning to her bed, she might be aroused again by an agitated Blackbird calling at the window for relief from a cat that threatened its brood. Again the tireless friend of birds would rush out to rout the marauder by dousing it with water. At dawn, or later in the day when sickness kept her in bed, her birds would pull her hair, or tug at her bedclothes, to remind her that they were hungry for the food she always had for them.

Pine Siskins of North America employed similar methods to stir a benefactor to action. E. R. Davis, a late riser, kept seeds for them in a

Blue Tit

glass-covered box on a windowsill beside his bed. Early in the morning the siskins would enter to claim his bounty. If he still slept or pretended to sleep, they would arouse him by pulling his hair, pinching his ears, tweaking his nose, even gingerly pulling up an eyelid to open an eye. If he took refuge under the covers, leaving open a tunnel-like passage for breathing, a venturesome siskin would by degrees advance into the tunnel until it reached Davis's face and tweaked his nose, which had the desired effect of making him jump up and uncover the tempting seeds.

In *Birds of the Grey Wind*, Edward Armstrong told of a European Robin who in winter came regularly to his enclosed porch to roost in a eucalyptus tree that grew there. The bird arrived early enough to be settled before the door was closed for the night, and in the morning he waited contentedly until the door was opened and he could fly out. This robin, who had never been fed or tamed in any way, continued his visits until someone tried to catch him. Unlike other birds that blundered into the porch and would have beaten themselves against the window panes had they not been promptly released, the robin never showed any sign of panic. Apparently, he had observed the daily opening and closing of the door and satisfied himself that it opened as regularly as it closed. All the foregoing birds acted intelligently, if sometimes inconsiderately, in situations for which their hereditary patterns of behavior contained no paradigm.

Birds appear to have an immune system, and their wounds and fractured bones heal quickly, but they can hardly have programmed methods for dealing with all the congenital or acquired deformities that sometimes afflict them. Birds with impediments are occasionally fed by their mates or companions, as has been reported of a Black-faced Wood-Swallow by Klaus Immelmann and a Black-headed Grosbeak by Wade Fox. More often, they keep themselves alive by a strong will to live and intelligent adaptation to their altered states. I watched a big Lineated Woodpecker whose strongly bent lower mandible crossed the upper mandible and projected so far to the left that at first sight he appeared to be carrying a long twig. With such a grotesque bill, he could not peck into wood or flake off bark. When I first saw him, he had already managed to live for a considerable period with his slowly developing deformity. He was laboriously extracting red arillate seeds from an opening pod of *Clusia rosea* and painfully working them back into his mouth; hanging back downward, he supported seeds on his foreneck until he could swallow them. He appeared to have adapted to a largely frugivorous diet, and for the next five weeks he frequently visited the *Clusia* tree. After that he vanished, probably having been unable to eat enough to survive a prolonged rain that left him looking bedraggled.

A Fiery-billed Aracari with an upper mandible so strongly bent upward and sideward that it could not close, was nevertheless in good plumage, possibly because flock mates supplemented the food it could gather for itself. A Chestnut-mandibled Toucan survived for at least two and a half years without the terminal half of its great upper mandible. It was alone whenever I saw it. Another Chestnut-mandibled Toucan, who had lost most of its lower mandible, was accompanied by a mate who may have fed it. A Speckled Tanager, with a right leg too severely injured to grasp a perch, managed to build a skimpy nest, lay the usual two eggs, and, aided by her mate and a helper, rear a normal fledgling. While standing on her nest, she supported herself by resting her half-spread right wing upon the rim.

One of Howard's Great Tits developed on the end of his upper mandible a spikelike projection that made him turn his head sideward to pick up food. Two days before his young were due to hatch, he began to work hard to break off this encumbrance by vigorous bill wiping, as though realizing that he could not feed nestlings while it remained. By the time they hatched, he had rid himself of the spike and was prepared to attend them. In a later year, this tit's abnormality grew anew and again he removed it just before his young were born. A female Great Tit removed a similar growth when her eggs were about to hatch. R. D. Manwell knew a Brown-headed Cowbird born with only one leg but who nevertheless managed to grow up, remain lively and healthy, and, apparently, to migrate at least once to Mexico and return to New York state. Survival with this congenital defect appears the more remarkable when we remember that cowbirds find much of their food on the ground. If we define intelligence as the ability to adapt to unusual conditions, all these birds who adjusted to their impediments certainly displayed it.

Whereas the field naturalist and the bird-loving householder seek evidence of fluid intelligence in behaviors that appear to be not innate but spontaneous innovations of free individuals, the experimental psychologist tends to equate intelligence with learning and tries to demonstrate it by animals' ability to solve puzzles devised for them. They may be set to run an intricate maze, with food at the end of the correct course. They may be taught to select a certain symbol or color among a range presented to them, with a reward for the correct choice. These puzzles can be made increasingly complex to test for degrees of mental ability. The number of trials, sometimes hundreds, needed to achieve proficiency are the measure of the animal's teachability or intelligence. The domestic pigeons and chickens most often used in these tests do about as well as, or better than, small mammals such as rats, rabbits, and cats but make lower scores than elephants, horses, donkeys, and zebras.

Since these laboratory tests demonstrate only some aspects of intelligence, the psychologists who conduct them are reluctant to rank the performers in the order of their mental capacities. Different lifestyles demand different aptitudes. What common denominator can serve for comparing the intelligence of a bird and that of a horse? The bird does many more things than the quadruped can do. Although a bird's complex behavior patterns are innate, it needs intelligence to adapt them to the endless details of a complex ambience.

It would be difficult for a laboratory experimenter to devise situations as complex as those that animals confront in natural environments. In the course of a year, a frugivorous bird encounters a great variety of fruits, differing in size, shape, color, texture, and taste or fragrance; some are edible and some inedible. Insects present even greater diversity to insectivores. It would be helpful to foraging birds to recognize fruits as a class that includes all its multiple varieties, and similarly for insects and other kinds of food. To avoid the many predators that menace it, a bird should be able to recognize them when at rest or in motion, from the front, side, or rear, in poor light as in bright; it should develop an image of each major kind of predator more fluid or flexible than that of a circle or a square. It should form recepts, or unnamed concepts, of categories or classes of things.

Docile pigeons were chosen by R. J. Herrnstein and his co-workers to test the birds' ability to learn open-ended categories similar to those faced in nature. In their first series of tests, as reported in a volume edited by Hoage and Goldman, they gave several pigeons the task of distinguishing people from anything else. They showed the birds twelve hundred slides, of which half contained one or more people while the other half had none. The humans might be children or adults, males or females, of various races in assorted poses, shown whole or in part, in the foreground or the background. The pigeons were rewarded with food for pecking at a but-

ton when pictures of people were shown, but received nothing for pecking when they saw photographs without people. By pecking at different
rates for different types of pictures, the pigeons seemed to reveal that they
had formed a category or recept of people, broad enough to encompass a
great diversity of them. Free birds have a similar recept; in regions where
they are persecuted by people, they recognize, and are wary of, humans
of whatever size, race, or sex, whatever they wear, and whether they see
people whole or in part.

Other studies showed that pigeons could form different categories.
They could recognize water as a drop, a puddle or pool, or an ocean; a tree
whether they saw it whole or in part; whether they were shown things on
slides that were familiar or new to them. To test the scope of pigeons' ability to categorize, they were required to distinguish the capital letter A, in
various type styles, from the number 2, and they succeeded. They could
even learn to distinguish pictures portraying fishes in various orientations
from pictures without fishes. Evidently, pigeons' ability to form categories
is not limited to things important to them in natural habitats—such as
trees that give shelter and sites for their nests, and people who might harm
them—but is much more general; and it can be cultivated by experi-
menters who reward them with food for categorizing things so remote
from their normal lives as printed symbols and fishes.

Often when one suddenly approaches a wild pigeons' nest, the incubating bird flies off so abruptly that it knocks an egg or two from the shallow bowl, giving the impression that it is a foolish bird. But watching from
a blind, as I have done with several species, one gets a very different impression. The sitting bird is alert to every sound and movement. If the
approaching animal is a cow or horse, the bird stays on its nest. If a
person—a hereditary enemy—approaches, the pigeon weighs the chances
of remaining undetected, for its departure will betray the nest. If, at the
last moment, it becomes evident to the pigeon that the human has noticed
it, the bird hastily departs, saving its life, often with the loss of one or both
of its eggs. The ability to delay response to a threat or an enticement while
awaiting developments, as pigeons and many other birds do, is an indication of intelligence. Unlike some other birds, a pair of pigeons can simultaneously attend a nestling that has fallen prematurely from a high nest
and one that remains in the nest.

Ornithologists are sometimes asked which birds are most intelligent.
An answer often given is crows, ravens, and related birds—the corvids.
These large, aggressive, opportunistic omnivores exhibit great behavioral
flexibility by taking foods so diverse as fruits, insects, small living vertebrates, carrion, and much else. When removed from the nest before they
are well feathered to be hand-raised, they become strongly attached to

their foster parents, often regarding one as a mate. Their tameness rec-
ommends them for the intelligence tests that experimenters give them in
laboratories, and they make relatively good scores. Their intelligence won
them a place at the summit of the evolutionary tree in certain older systems of classification, although now, as in the most recent check-list of
the American Ornithologists' Union, they are placed near the bottom of
the Oscine passerines, with finches, weavers, and allied families at the top.

The great difficulty of sharply separating learned or innovative behavior, on the one hand, and innate or genetically determined behavior, on
the other, and the vast diversity of the lifestyles and activities of birds,
make it impossible to decide which are most intelligent. A large segment
of avian behavior, especially its more complex forms, is perfected by learning and experience building upon the innate foundation that we call instinct. Starting with an imperfect hereditary pattern of its species' song, a
songbird improves his performance by listening to his elders. Birds appear
to have an innate pattern of their nests; but at least the more elaborate of
them, such as those of certain African weavers, are not finished without
practice. Many studies have demonstrated that experience makes birds
more efficient parents; pairs nesting for the first time rear fewer young
than do older breeders. Although the impulse to fly in a certain direction
is innate in at least some migratory birds, the competent navigation that
many display by commuting annually between familiar winter and breeding territories, separated by thousands of miles, is not attained without
observation, learning, and experience. These are only a few of birds' activities in which learning complements innate tendencies.

In view of the difficulty or impossibility of drawing a sharp line between
innate and learned behavior, it seems improper to assess a bird's total mental capacity solely by performances that we can, with more or less confidence, categorize as learned or original. To judge it fairly, we should
survey all aspects of its behavior, innate and learned, which might make
comparisons difficult. Which is smarter, the Old World warbler who develops a repertoire of scores of songs or the New World ovenbird (Furnari-
idae) who builds and carefully maintains a complex nest of interwoven
twigs? The Woodpecker Finch of the Galápagos Islands who forages with
a tool or the cooperative breeder who adapts to the social complexities
of a closely knit group of a dozen individuals of both sexes and all ages?
The Jackdaw who makes a high score on a laboratory test or the migrant
finding last year's territory at the end of a journey of thousands of miles
over land and water? We must abandon the effort to discover which bird
is "most intelligent."

Nevertheless, it is doubtless true that species of birds differ in intelligence or mental capacity, as do individuals of a single species. On the

whole, foraging generalists, who take a wide variety of foods in diverse ways, seem to have minds more flexible and capable than do specialists whose diet is limited. Likewise, social birds—especially those living in co-operative groups in which the status and functions of individuals differ rather than those gathering in unorganized flocks—appear to be more intelligent than birds who live singly when not breeding. However, I believe that no bird could long survive in a complex, competitive world without a measure of intelligence, which might be difficult to demonstrate in a bewildered, depressed captive in a laboratory.

The mental development of animals is related to what they can do with their limbs. Because their wings carry them farther and faster than plodding bipeds can go, birds have a superior sense of location and direction and can perform feats of navigation that humans could not accomplish without instruments. But with a pair of hands equipped with opposable thumbs, the finest executive organs in the animal kingdom, people can manipulate things as no bird can do with its bill, admirably versatile organ that this is. Ancient thinkers knew that humans are the most intelligent animals because we have supple hands.

To have a mind that can think of many things (such as building safer nests for eggs and young or more comfortable dormitories for sleeping) that it cannot do because of the limitations of its bill would be of little practical advantage to a bird. On the contrary, to be unable to execute brilliant ideas because of the inadequacy of limbs might lead to a debilitating sense of frustration, as occasionally happens to us when we try to perform a seemingly easy operation for which our hands prove inadequate. Birds appear to have as much mental capacity, call it instinct or call it intelligence, as would be useful to them. To have more might be upsetting.

References: Armstrong 1940; Davis 1926; Fox 1952; Herrnstein et al. (from next citation); Hoage and Goldman 1986; Howard 1952; Immelmann 1966; Manwell 1964; Skutch 1980.

Chapter 12

❦

Apparently Stupid Behavior

A TINY Blue-crowned Manakin from the neighboring rain forest flew into a room of our house in Costa Rica. We opened wide the windows and door leading outside to facilitate the bird's escape. Ignoring these avenues to freedom, he continued for many minutes to fly around beneath the high ceiling, clinging to the walls here and there, always above the windows and doorway. We tried to help him find an opening by gently pushing him down with a broom, but he stubbornly persisted in remaining too high. Finally, exhausted by his vain efforts to escape, he sank to the level of a window and flew away.

Small flycatchers, tanagers, and most hummingbirds that enter the house behave in the same exasperating manner, remaining above our reach and frustrating our efforts to release them before they hurt themselves or stain walls and furnishings with their droppings. Their failure to take advantage of wide-open windows a few feet below them seems utterly stupid until we reflect that all their lives, and through those of their ancestors for countless generations, the foliage above them was rarely so dense that when in trouble they could not fly upward through it. Habit blinded them to the obvious, as it sometimes does with us. An exception to such behavior is the Little Hermit hummingbird, who frequents the undergrowth of woodland and thicket, rarely rising high. When it enters a building in search of spiders, it readily finds an open window or door, or in a screened room it hovers along the hardware cloth until it finds an exit.

On rural roads, one often sees domestic chickens dashing across in front of a speeding car. Having been surprised by an advancing vehicle while foraging on the far side of the road from home and ignoring the fact that they would be safer on either side than in the middle, in a panic they expose themselves to the danger of being struck. Like many another creature, they feel more secure where they belong. Their behavior seems stupid but is not without a reason.

In a cloister around the patio of a hotel in Antigua, Guatemala, I watched a Rufous-collared Sparrow persistently fighting his reflection in a mirror—behavior that has frequently been reported of birds in other lands. Brown-headed Cowbirds visiting Hervey Brackbill's windowsill feeding shelf in the suburbs of Baltimore displayed to their images in the pane and some repeatedly attacked them. One male cowbird continued for over two months to peck at and bow to his reflection. Can't the foolish creatures see that they are sparring with their images reflected from the glass?

But consider the superstitions about reflected images among many peoples and the confused explanations of vision by ancient and not so ancient philosophers. In *The Golden Bough*, Sir James Frazer records how a tribesman, viewing his reflection in clear, still water or in a mirror supplied by a trader believed that his external soul was returning his gaze. In crocodile-infested waters, he needed to be careful lest one of these reptiles seize his soul and drag it beneath the surface, thereby causing his death. The Classical myth of the beautiful Narcissus, who languished and died from viewing his reflection in a woodland pool, reveals the persistence of such belief in ancient Greece. In more recent times, the widespread habit of covering mirrors or turning them to the wall when a member of the household dies springs from the fear that the ghost of the deceased may seize the soul of a survivor, projected from body to mirror, and carry it away.

Classical philosophers, as exemplified by Plato in his cosmological dialogue *The Timaeus,* taught that a visual ray streaming outward through the pupils of the eyes, coalesced with light coming from an object to bring to the viewer the object's image. As every educated person now knows, nothing issues from the eyes to give us vision; the incoming light rays are all that we need for sight. When we recall all the confusions that have surrounded reflections, and vision in general, until relatively recent times, it should not be surprising that a bird is deceived by its image in a mirror or in the shiny hubcap of a car, mistaking this for a rival who must be expelled from his territory—much as birds are confused by hearing their own voices reproduced on a tape recorder such as bird watchers use to draw them closer.

A few years ago, we reared a Gray-headed Chachalaca with the help of a domestic hen, having taken the egg from an abandoned nest. The first time that Laca saw herself in a mirror, she pecked at her image, which of course returned the peck, bill meeting bill repeatedly on the glass. Evidently, Laca did not recognize herself. After a few more experiences with

Gray-headed Chachalaca

mirrors, she looked at her reflection without pecking it. Had she become aware of the deception?

The unconfined Great and Blue tits who had the freedom of Len Howard's cottage paid little attention to mirrors. On first seeing themselves in one, they looked behind it, found no bird, and thereafter ignored reflected images. Howard noticed only one exception to this.

With prolonged, expert coaching, pigeons can be trained to peck at spots on their own plumage that they can see only in a mirror. They appear to realize that they are looking at images of themselves. Domestic chickens are apparently unable to learn that mirrors reflect themselves. When first introduced to a mirror, chimpanzees mistake their image for another individual, but after a few days' familiarity with mirrors, they behave as though they recognized themselves, touching parts of their bodies while looking at the glass. Orangutans evince the same awareness. Tested in the same way, gorillas and several kinds of monkeys appear unable to learn to associate their mirror image with themselves. When about ten months old, human children recognize themselves in mirrors.

A completely isolated female domestic pigeon will not lay an egg if provided with a nest. However, she will lay if she can see herself in a mirror placed in her cage; the effect is the same as when two females are confined together, or when a female is separated by a clear glass partition from another female or a male in an adjacent cage. Although in all these situations a mature female will eventually lay, she lays much more promptly when given a partner of the opposite sex, as L. Harrison Matthews demonstrated. Evidently, the female mistakes her mirror image for a living companion.

Window glass confuses birds. Since in the wild they can always fly directly to what they can clearly see, it is not surprising that when trees or shrubs are visible through two windows on different walls of a room, birds do not hesitate to fly toward the plants, not noticing the transparent, unyielding barrier against which they crash, too often killing themselves. A recent estimate by D. Klem, Jr., places the number of casualties at windows in the United States and Canada at one hundred million to one billion annually. Screens of fine hardware cloth that hardly diminish visibility likewise deceive birds. Hummingbirds ram their slender bills into screens so far that they cannot pull free, with the result that they hang there and succumb to desiccation unless promptly released. Even a large toucan, with much greater impetus, may drive its great beak through a porch screen so deeply that it cannot extricate itself. If people who live surrounded by birds would drape or otherwise cover or place markers on at least one of every two glazed or screened windows or doors through which their feathered neighbors can see foliage, this distressing mortality would be greatly diminished. Rare is the bird whose careful parent teaches it the perils of window glass, as Howard's Great Tit, Jane, did.

Other unfamiliar materials may baffle birds, too. Behind our house stands a chicken roost, a tall, roofed structure supported on four posts, each surrounded by a broad metal band to prevent predatory quadrupeds from climbing up. As an additional safeguard against jumping animals, on the uphill side of this contrivance I fastened broad, thin sheets of metal, which rusted to a color not greatly different from that of some decaying trunks.

This chickens' shelter was the scene of the most fantastic performance that I have witnessed in birds. A pair of Slaty-tailed Trogons had come from the nearby forest to seek a nest site in our garden. First, they tried to dig a cavity for their eggs in the mass of decaying vegetable debris at the base of the spreading crown of a tall African oil palm. When this location proved unsatisfactory, they examined a decaying stub on the hillside behind the house, finding it also unsuitable. Then they turned their attention to the chickens' shelter, flying against the sheet metal from a neighboring tree with a *bang* that resounded through the house.

For two days they tried to make a hole in this vertical sheet that would have led to the open air on the inner side—apparently, they never looked behind it. One afternoon they continued for nearly four hours to fly noisily against it, striking it with their underparts—chest, feet, or both—as though to cling there and start the excavation. In one hour, the slaty

Slaty-tailed Trogon

female trogon struck the metal thirty-two times; her brilliantly green, red-bellied mate, nine times. On other occasions he was the more active partner in the futile undertaking. Finally convinced that they could not penetrate the metal, they returned to the forest, where in a massive stub with soft, decaying sapwood they started an excavation that was thwarted by the harder heartwood. After all these failures, they flew beyond my ken.

The short, thick bills of trogons are less efficient for carving the cavities in which they rear their broods than are the sharp bills of woodpeckers. Trogons do better when they dig into arboreal termites' nests, composed of hard, thin, curving sheets that they seize in their bills and wrench away, but such structures are not always available. With varying success, they try to carve chambers in trunks, which are sometimes so decayed that they crumble, whereas others have a deceptively workable exterior masking a hard core that foils their efforts. Often they must try again and again to finish a nest cavity in wood. This situation has favored an exploratory approach to their problem. I have known New World trogons of various species to nest in sites so diverse as decaying stubs, termitaries, papery wasps' nests high in trees, compactly massed roots of epiphytic ferns, and once (a Violaceous Trogon) in a high, earthy ants' nest attached like a stalactite beneath a lofty bough—the only record of a bird breeding in an ants' nest in the New World that I know. The Slaty-tailed Trogons' determined attack on the rusty sheet metal appeared to be an instance of this exploratory habit.

Birds can be deceived by more systematic means than the unintended effects of our own constructions. A small songbird mothering a nestling European Cuckoo while her own hatchling, heaved out of the nest on the concave back of the intruder, lies cold and starving below, is the classic example of the blindness of innate behavior. Can't the parent see that her own progeny needs her attention? Equally unperceptive are the many birds in the western hemisphere that hatch the intruded eggs of the several species of cowbirds and feed the ravenous fosterlings until they are bigger than the parents' own young. These are only a few of the world's many species of obligate nest parasites, but they are the most intensively studied, with a vast literature about them.

Although European Cuckoos and American cowbirds have much in common, and present the same problems to their victims, they differ in important ways. The cuckoo's eggs mimic those of its hosts, undoubtedly the consequence of a long evolutionary history; cowbirds' eggs show no tendency to resemble those of hosts that they have more recently parasitized. By an incredible infantine effort, the hatchling cuckoo heaves from the nest everything that it finds beside itself; only one European Cuckoo is reared in a nest. Nestling cowbirds have no such habit; one or more may grow with one or more of the host's young. Because female cowbirds, like female cuckoos, remove eggs from the nests they invade, and because the young parasites demand much food that might be given to the legitimate progeny, each young cowbird is, on average, reared at the cost of about one of the host's nestlings.

Before accusing the cuckoos' and cowbirds' hosts of incompetence and negligence, let us see what they do to prevent the violation of their nests. In the first place, they try to drive the adult parasites away from their nests, scolding, threatening, and striking them. The more heavily birds are parasitized, the more vigorous their efforts to prevent intrusion. Neighboring birds sometimes come to the aid of the defenders, as when Kenneth W. Prescott watched a male Scarlet Tanager join a pair of Red-eyed Vireos in attacking a Brown-headed Cowbird. Early in the morning, the bold parasite entered the nest and laid her egg while the defenders continued to buffet her with their wings. As is usual with cowbirds and cuckoos, she deposited her egg in less than a minute, not delaying in the nest for many minutes, as females laying in their own nests frequently do.

Many birds soon rid their nests of the intruded eggs. Those with gapes wide enough simply pick up the eggs and toss or carry them away. Birds with smaller but sharp bills drive their mandibles through the shells and

European Cuckoo (fledgling, left) and Reed Warbler

throw the eggs out. This is not without danger to the host's eggs because both cowbirds' and cuckoos' eggs have thicker, harder shells than those of other birds of similar size, which may deflect blows aimed at them to pierce the hosts' eggs lying in contact with them. Small birds unable to remove the foreign eggs may simply abandon their nests and build anew, if the season is not too far advanced. Less often a small bird isolates a cowbird's egg by building a new nest on top of it, a process that may be repeated if the cowbirds' intrusions continue, until the harassed parent has a nest four or five stories high, as occasionally happens with the Yellow Warbler. Among North American birds that regularly eject cowbirds' eggs are Eastern Kingbirds, Blue Jays, Gray Catbirds, Brown Thrashers, American Robins, Cedar Waxwings, and Northern or Baltimore Orioles.

European birds that toss out cuckoos' eggs include Fieldfares, Song Thrushes, Eurasian Blackbirds, Redwings, Spotted Flycatchers, and Bluethroats. As would be expected from the more disastrous effects of harboring a cuckoo's nestling than a cowbird's nestling, small-billed birds that puncture-eject cuckoos' eggs are relatively more numerous than those that puncture cowbirds' eggs, as A. Moksnes and his co-workers showed.

Despite all precautions, many birds are burdened with cuckoos' eggs and hatch them. The parents find in their nest an alien which they may not recognize as such, especially if they are young and inexperienced, while one or more of their own nestlings lie below it. What can they do for the dying chick? They may not be able to pick it up and replace it in the nest without injuring it in their bills. If they succeeded, the young cuckoo would heave it out again, and, until their own youngster has fledged, they cannot properly attend it except in the nest. One of the least resistible of all the impulses of altricial parents is to place food in the gaping mouth of any nestling they find in their own nest—and often in other birds' nests.

Evolution has played them one of its cruel tricks. I think we can best sympathize with the predicament of a bird that finds in its nest a nestling different from its own if we remember what people do when they have the misfortune to become parents of an incurably afflicted baby. Restrained by mother love from permitting the unfortunate baby to fade quietly away, (and prevented by law from actions once sanctioned in similar circumstances), they undertake to rear a youngster who requires more care than several normal children and may bring them little joy.

When we survey the situations in which we find birds behaving with apparent stupidity, we find that they are unnatural, created relatively recently by people, or are difficult predicaments caused by highly evolved nest parasites that exploit other birds' strong parental instincts. Not infrequently, the birds are smart enough to detect deception or avoid unnatural perils. When we find a bird acting ineptly, it would be valuable to know its age. Human children tend to behave foolishly more frequently than adults do. May not the same hold true of birds?

References: Brackbill 1961; Cornford 1937; Frazer 1944; Friedmann and Kiff 1985; Hoage and Goldman 1986; Howard 1952; Klem 1990; Matthews 1939; Moksnes et al. 1991; Plato (see Cornford); Prescott 1947.

Chapter 13

❧

Freedom
and Altruism

TO BE "free as a bird," untrammeled by petty cares, able to fly lightly over barriers and chasms that defy our plodding footsteps, is a way of living for which we sometimes yearn. Where external obstacles are absent, internal control may limit freedom. A bird's life is controlled by a hereditary pattern which is probably not felt as repressive or restrictive because it is not externally imposed. Moreover, the pattern is loosely articulated to permit its adjustment to variable external circumstances; without a measure of freedom, a bird could hardly survive and reproduce. Birds demonstrate their release from strict control by their genes in at least six ways.

1. Birds with large repertoires vary the order of their songs. Mimetic birds select from the sounds they hear those to imitate and reproduce them in diverse sequences, as suits their fancy. Medley singers jumble their notes.

2. Female birds freely choose their partners in reproduction. In a courtship assembly where many males try to entice them, they bypass eligible suitors with whom they would mate if acceptance were a reflex act, to delay until they find the individual that impresses them most favorably, probably after comparing a number of rivals. Among birds that breed in pairs, a female appears to assess both the male who sings to attract her and the territory he offers. Sometimes, as in Bobolinks and Dickcissels, she prefers to become the second spouse of a male with a superior territory rather than the only mate of a male with an inferior domain, as though realizing that her nestlings will be better nourished with little or no help from a bigamous father on a territory that produces abundant food than by a monogamous father who gives them his full attention on an unproductive territory.

3. Among cooperative breeders, nonbreeding helpers, usually progeny of the breeding pair, may remain to attend their younger siblings for only one year or, rarely, for as many as four or five. When a helper leaves its group to search for one in which it may become a breeder but fails, it may return and be accepted by its natal group. It freely chooses the course it takes.

4. The type of nest a bird builds, whether open or closed, low or high on a bank, or in a cavity of some sort, is determined by its heredity, but too narrow specification of the site might often prevent breeding. Reproduction is promoted by a flexible program that permits free choice of nest sites within more or less wide limits. Common Bush-Tanagers and Flame-throated Warblers may nest high in trees or on the ground. Violaceous Trogons carve chambers for their eggs in vespiaries, ants' nests, decaying trunks, or amid tightly massed roots of epiphytic ferns. Male Marsh Wrens and Winter or Old World Wrens start a number of nests, one of which is freely chosen by the female who lines it and lays her eggs in it.

5. Birds demonstrate their freedom by playing. I doubt that many of the games described in chapter 5 have enough survival value to be fostered by natural selection and programmed in the players' genes. Birds have sufficient opportunity to exercise their limbs and strengthen their muscles in daily, life-supporting activities without indulging in seldom-seen frolics. Especially when birds take advantage of some rare opportunity for playing, as by sliding down a smooth, slippery incline or riding rapids, it is highly improbable that their behavior is specifically determined by their genes. In choosing their games, birds manifest their freedom and mental flexibility as well as their capacity for enjoyment.

6. Helpers are birds that serve others who are neither their mates nor dependent young. In cooperative breeding, helping is part of a hereditary lifestyle that binds parents and progeny into a mutually supportive group, but more sporadic helping occurs apart from this familial context. Birds of various kinds feed or otherwise attend young of neighboring pairs, of their own or different species. Interspecific helping is more likely to be noticed than intraspecific helping because two kinds of birds bringing food to the same nest attract the attention of even casual watchers, whereas to learn that more than two of the same species are feeding a brood requires more careful attention. Moreover, among territorial birds, nests of different species are more often close together than are two nests of the same species.

Among the many cases of interspecific helping recorded in my *Helpers at Birds' Nests* are the following:

A Gray Catbird fed and mothered an orphaned brood of Northern Cardinals. A House-Wren gave food to both parent Black-headed Grosbeaks, who ate some and passed the remainder to their nestlings. After the young grosbeaks fledged, the wren fed them directly. A few days later, this wren brought food to a family of House Sparrows. A Swainson's Thrush assisted in feeding nestling American Robins. A Black-and-White Warbler repeatedly fed nestling Worm-eating Warblers, against the opposition of their parents. A House Sparrow brought food to three fledgling Eastern Kingbirds. A Scarlet Tanager fed young Chipping Sparrows. A male Red-legged Honeycreeper repeatedly fed a fledgling Scarlet-rumped Tanager twice his size. These are only a few randomly chosen cases of interspecific helping. When I reflect upon the number of such events that have been reported, and the infinitesimally small proportion of birds that humans have watched, I suspect that every species of bird occasionally helps every other species of similar size and habits that nests nearby.

Male birds with too little to do while they wait for their mates to hatch their nestlings sometimes feed young of a neighboring family, often of a different species. Parents who have lost their own brood sometimes satisfy thwarted parental impulses by attending neighbors' young. Or, after rearing to independence their own progeny without exhausting their parental fervor, they turn their attention to other birds' offspring. Sometimes the sight of a nestling's gaping mouth elicits a meal from a parent carrying food to its own brood. To feed offspring or mates is one of the strongest impulses of altricial birds, on which the perpetuation of their species depends. Although the same genetic code that impels the act directs it to the bird's own progeny or mate, occasionally birds reveal their freedom from strict control by their genes by feeding young of different parents, of their own or another species.

Holding that an animal should devote all the energy it can spare from self-maintenance to the propagation of its own genes, biologists regard helping, especially when interspecific, as a mistake. It is not always clear in what sense they regard it as an error: whether the helpful bird misuses its energy, or whether it confuses its beneficiary with its own progeny. I doubt that either the Gray Catbird, whose own nest is an open cup in a bush, was unaware that the House-Wren nestlings it fed in a birdhouse were not its own species, or that the House Sparrow feeding fledgling Eastern Kingbirds mistook them for young sparrows. Birds that can recognize individuals of their own species by differences in appearance so slight that they escape keen human observers are not likely to confuse another species with their own.

Neither can we regard interspecific helping as widely detrimental to a bird's own reproduction. Occasionally this does occur, as when a female Tropical Gnatcatcher, fascinated by a brood of Golden-hooded Tanagers, fed them regularly, ignoring the nest her mate was building in a neigh-

Tropical Gnatcatcher (female, left) and Golden-hooded Tanager

boring tree. Much more frequently, the helpful bird does not diminish its own reproduction, as when a male feeds neighboring nestlings while waiting for his own to hatch, or when parents who have lost their nest feed a neighbor's brood, probably when it is too late in the season for them to renest. Although widespread, interspecific helping is too rare to be of evolutionary importance; it claims our attention as an indication of freedom and altruism.

Altruism is the conferring of benefits, with no reward beyond the satisfaction that beneficence gives to the benefactor. With us altruism is a virtue whose moral value depends upon its intention. Since birds' minds are closed to us, we cannot read their intentions; but objectively the bird who feeds another's progeny is an altruist. Biologists, who define altruism in a special way, as an organism's promotion of the reproduction of others at the cost of its own, are perplexed by it because, contributing nothing to the survival of the altruist's species, it would not be fostered by natural selection.

Altruism becomes understandable when we compare it to play, or to artistic creation not motivated by gain. All three are the application of energy in excess of vital needs to intrinsically satisfying activities. The altruistic bird satisfies a strong impulse by helping to rear alien young when its own do not claim all its energy. Sometimes the helper is so strongly attracted by neighbors' nestlings that it feeds them against the opposition of parents who resent this uninvited intrusion; but often parents and helper cooperate amicably to nourish the brood. Occasionally two pairs of par-

Field Sparrow (left) and Rufous-sided Towhee

ents work together to rear two broods in neighboring nests, each feeding the other's young as well as its own, as when Rufous-sided Towhees and Field Sparrows jointly reared their broods—a case of reciprocal altruism.

The freedom that I attribute to birds is not classical free will, according to which a free volition is an uncaused motive of action, an instance of indeterminacy. Not only do we lack evidence for the kind of free will demanded by some moralists, but it would abrogate moral responsibility; we could not be held accountable for acts undetermined by the constants of our character. The foundation of our freedom (as I pointed out in *Life Ascending*) is our ability to delay action while we mentally explore alternative possibilities, choosing the course most attractive to us. When passion precipitates action without foresight, freedom is lost. A bird selecting a mate, territory, or nest site evidently delays action while she examines alternatives, thereby exercising free choice. Thus, we find in birds freedom on two levels: release from strict genetic control, as when they play, diversify their singing, or feed unrelated young; and at the higher level, while they delay to choose a sexual partner or a territory until they have viewed alternatives. Not only because it can fly is a bird free.

References: Hartshorne 1973; Shy 1982; Skutch 1961, 1985, 1987b, 1992.

Chapter 14

❧

The Brain and Senses

THE BRAIN is the seat of the mind, the organ by which mind interacts with body and, through the senses and muscles, with the surrounding world. Because birds have small brains, and because most people know birds so superficially, they are widely believed to have little intelligence. In a popular book on the human brain, G. N. Ridley wrote: "The bird has ingrained, natural impulses to do certain things in a stereotyped way, and such behaviour patterns are operated, not through the medium of the cortical cells, but through the nervous connections in the striped bodies. The bird's brain is a beautifully adjusted instinctive machine, its body specialized for the performance of routine duties, and in consequence the creature has little or no ability to experiment and to modify its way of life." This appraisal is more complimentary to birds than others that one might quote.

Built on the basic vertebrate plan, the brain and nervous system of a bird closely resemble those of reptiles, especially crocodiles and certain lizards, but a bird's brain is ten or more times larger than that of a reptile of similar body size. In both birds and mammals, the size of the brain increases with that of the body. The average brain of an adult human weighs about 46 ounces (1,300 g), slightly more in males than in females. An elephant's brain weighs about 11 pounds (5 kg), and that of one of the great whales about 22 pounds (10 kg). However, the brain's weight does not increase proportionally to that of the body. The whale has one gram of brain for 8,500 grams of body, whereas humans have one gram of brain for 44 grams of body. The small New World monkeys called marmosets, with one gram of brain for 27 grams of body, have relatively much bigger brains than do humans, but in this respect they are surpassed by capuchin monkeys, with one gram of brain for only 17.5 grams of body. Not surprisingly, capuchins are among the most intelligent of nonhuman primates, rivaling the great apes. With bodies kept light for flying, birds have small brains. Nevertheless, their brains tend to be as massive as those of mammals of the same body weight. In a comparison of the two great classes of warm-blooded vertebrates, birds are not inferior.

Brains also differ in the wrinkling or convolutions of the cerebral hemispheres, which increases their surface where the sensory and motor centers are situated, and where the "gray matter" with which we supposedly think is concentrated. The cerebral cortex of a whale is more densely wrinkled than that of people or chimpanzees, whereas that of capuchin monkeys has only a few folds on either side. The convolutions of the two hemispheres of a human brain may show quite different patterns. In contrast to certain mammals, the surfaces of birds' cerebral hemispheres are unwrinkled. However, the external striatum, just inside the smooth cortex, compensates for the poor development of the latter and might be grouped with it, for it receives sensory information from eyes, ears, and, to a lesser degree, from tactile stimulation, through the thalamus, as in mammals. Part of the external striatum appears to correspond to the mammal's motor cortex because it has direct nervous connections with the spinal cord. Despite great differences in the sizes of brains and the relative proportions of their regions, those of all multicellular animals are composed of similar neurons and synapses, their working parts.

Reviewing all the evidence, especially with reference to mammals although it should also apply to birds, Franz Weidenreich concluded that "neither the size nor the form of the brain or the surface of the hemispheres or the wrinkle pattern in general or in detail furnishes a reliable clue to the amount and degree of general or special mental qualities." However, particular regions of the avian brain are developed in relation to their special senses. Since birds' eyes are their chief sources of information, their optic lobes, at the sides of the cerebellum, are large and prominent. Because birds depend much less upon the sense of smell, the olfactory lobe, at the front of the brain, tends to be much smaller than in mammals, to which odors are more important.

Behind birds' cerebral hemispheres lies the cerebellum, which has large neural connections with the centers in the spinal cord that control posture and equilibrium. In flight it acts like an "automatic pilot," maintaining stability and keeping the bird on its course. The cerebellum of birds is not, like that of mammals, divided into hemispheres, possibly because birds manipulate objects with a single bill, whereas mammals employ two hands or forepaws.

The mind receives the information that helps direct its interactions with the outside world, and in fortunate circumstances adds to the enjoy-

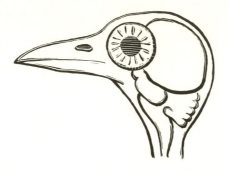

Bird's brain and eye

ment of life, through the senses that report to the brain. In general, the bird's senses resemble ours. Like us, they are primarily diurnal animals whose most important sense is vision. This is reflected by the size of their eyes, which are about as large as their heads can accommodate. Eyes on opposite sides are almost in contact inside the head, and together they may weigh almost as much as the brain. Unlike our eyes, those of most birds, with the notable exception of owls, are directed sideward instead of forward, which gives them a wider field of vision for detecting enemies. Because they are less spherical and more tightly enclosed in the head, birds' eyes are less mobile in their orbits than human eyes are.

Although birds cannot focus both eyes on a small object in front of them as well as we can, for flying they need to see straight ahead, and the visual fields of the two eyes do overlap in all species, from ten to thirty degrees in most granivorous birds to thirty-five to fifty degrees in insectivores and hawks. Birds' highly flexible necks compensate for limited binocular vision; to examine a nearby object, they often turn the head sideward and scrutinize something with one eye. The thin white nictitating membrane, which from the inner corner moves back and forth across the eye, serves to clean, moisten, and protect the cornea.

The structure of the avian eye is basically the same as that of the mammalian, with a cornea, iris, and a lens that focuses light upon a retina covered with rods and cones. The iris, which may be black, brown, red, yellow, blue, or nearly white, occupies the whole exposed part of the eye and is readily contracted, increasing its own prominence while narrowing the pupil. Unlike our eyes, those of many birds have two foveae, or spots of acute vision, one in the center of the retina for monocular sideward vision, and one toward the outer side, for binocular frontal vision. In these shallow pits the cones are extremely fine and tightly crowded, well provided

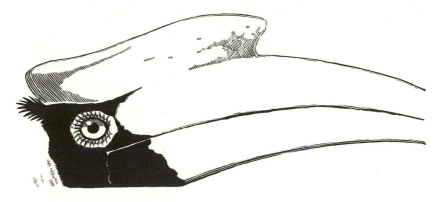

Great Hornbill's eye

with nerves for conveying fine details to the brain, and without the over-lying layers of retina that interfere with clearest vision.

A peculiar feature of the avian eye is the pecten, a pleated or veined structure projecting into the chamber behind the lens, where its shadow falls upon the blind spot in the retina. It is composed almost wholly of blood vessels and pigmented connective cells. Smallest in nocturnal birds, larger in seedeaters, it is best developed in insectivorous species and diurnal raptors. Of the several theories about the pecten's function, the most probable is that it supplies, by osmosis, nutrients and oxygen to the retina, which is devoid of blood vessels that would interfere with vision.

An eye's visual acuity depends upon its size, structure, and the fineness of the retina. By packing visual cells, especially cones, more densely over the retina, little birds compensate for the smallness of their eyes. In birds, as in ourselves, three pigments, for red, green, and blue, present in the cones, are the basis for trichromatic color vision. In addition to these pigments that birds share with certain mammals, the cones of diurnal species contain brightly colored oil droplets that are absent from the rods. At least five types of droplets, diversely colored with carotenoid pigments, are variously distributed over the retina. By shielding the visual pigments from certain wave lengths, they increase the effectiveness of others. Not only can birds see all the colors that we perceive, but the vision of pigeons, certain humming-birds, and probably many other birds that have not been tested, extends into the ultraviolet, to wave lengths between 325 and 360 nanometers, to which we are blind as our range of vision does not extend beyond 400 nanometers.

Polarized light consists of waves that vibrate in a single plane, instead of all planes, as in light from the sun or an electric bulb. At least some birds, of which pigeons are the best-known examples, can detect the plane

Scissor-tailed Flycatcher

of polarized light from the sky, which is determined by the Sun's position and may serve birds for orientation when the Sun is not directly visible. With their preponderance of cones for color vision, the retinas of diurnal birds are poorly provided with rods for seeing in dim light. Nevertheless, pigeons and domestic chickens can adapt to nocturnal darkness about as well as we can, although the process is much slower, taking about an hour instead of ten minutes. This ability may save the lives of purely diurnal birds by enabling them to find shelter if frightened from their roosts on nights that are not too dark, and it must be helpful to the many nocturnal migrants.

The ears of birds are inconspicuous because they lack the pinnas or projecting lobes that in humans and other mammals more or less effectively catch the sound waves in the manner of ear trumpets formerly used by people with poor hearing. Moreover, birds' ears are hidden by feathers, the ear coverts. Otherwise, birds' ears are not greatly different from ours and serve the double function of hearing and preserving the body's balance by means of three semicircular, liquid-filled tubes in three different planes, all at right angles to one another. Within the range of wave frequencies that birds share with people (1–5 kHz), they are about as sensitive as we are to changes in the frequency, duration, and intensity of simple pure tones. Above this range, their sensitivity decreases rapidly, but some can hear frequencies up to about 10 kHz, to which most of us are deaf. Like mammals, birds can localize the sources of sounds. Like bats, a few birds can guide their flight in complete darkness by emitting sounds and listening to their echoes. The notes that birds use for this purpose are mostly within the range of human audition, whereas those of bats are too high-pitched for us to hear. Ecolocation is practiced by Oilbirds, who rest by day and nest in dark caves of northern South America and Trinidad, and by those cave swiftlets that nest in the darker reaches of caverns in southeastern Asia and Indonesia.

Although birds have well-developed olfactory organs in their paired nasal cavities, it is commonly believed that their sense of smell is weak or lacking. Whereas stalkers of mammals try to approach them upwind, so that the breeze does not waft human scent to the sensitive nostrils of wary creatures, bird-watchers pay little attention to wind direction, no matter how carefully they conceal themselves from birds' keen sight. Sophisticated modern procedures in the laboratory have demonstrated that pigeons, chickens, ducks, and other familiar birds can indeed detect a diversity of odors, but only in a few species does olfaction appear to play an important role in birds' lives. One of these is the Kiwi of New Zealand, a flightless bird that hunts by night, probing loose soil and ground litter with its long, slender bill, and detecting its small invertebrate prey by smell. Relative to its size, the Kiwi has at the front of its brain one of the largest olfactory lobes in birds. Leach's Petrels find their way to their island nesting burrows on dark nights by following upwind the distinctive odor of each. They also have large olfactory bulbs, as do other petrels and shearwaters, who also probably find their nests by scent. Even albatrosses, belonging to the same order of tube-nosed swimmers, have large olfactory bulbs, although they nest in the open instead of in burrows. Like other tube-noses, they find food by its scent. To judge by the size of their bulbs, aquatic birds in general depend more upon smell than do terrestrial birds.

The history of ornithology contains a long controversy over how vultures find their food, some distinguished ornithologists maintaining that they use their noses; others, their eyes. The confusion resulted not only from crude methods of investigation but from failure to realize that different species may have different means of detection. It remained for Kenneth Stager to demonstrate conclusively that, among New World vultures, the Turkey Vulture hunts for carrion with its nose whereas the Black Vulture, who at least in the tropics includes much fruit in its diet, uses its eyes to find food. Correspondingly, the olfactory lobe of the Turkey Vulture is much larger than that of the Black. Also large are the olfactory lobes of flightless ratite birds, the Ostrich, Emu, and Rhea.

Homing pigeons, who have relatively large olfactory lobes, may use odors to help them find their way, at least within moderate distances of up to about sixty miles (100 kilometers). Winds carry to their lofts diverse scents from different directions, by means of which they develop an "odorous map," or mosaic of smells of the surrounding territory. They associate one scent with a certain direction, a different odor with another direction, and this information helps guide their homeward flight when they are released at a distance. This hypothesis is still controversial; apparently, use of olfaction for orientation varies with the strain of pigeons, their training, and the region where they are tested.

As with smell, birds are commonly supposed to have a poor sense of taste. Some, including domesticated breeds, gobble down their food so rapidly that they seem hardly able to savor it. However, birds, with pulse and breathing rates much higher than ours, live at a swifter pace than we do and may taste the food that they so quickly swallow. Frugivorous birds who mandibulate berries to remove their skins, and seedeaters who crush seeds in powerful bills before swallowing them, certainly have time to taste their foods. When pigeons encounter an unfamiliar food, they repeatedly pick it up and drop it before swallowing or finally rejecting it, apparently judging its edibility by taste. After one unpleasant or distressing experience with an unpalatable insect, a bird thereafter eschews it—a caution that makes warning coloration and mimicry profitable to butterflies and other insects.

Suggested by field observations, birds' ability to discriminate tastes is corroborated by laboratory tests, which demonstrate their sensitivity to salts, sugars, and other substances. During their first days after hatching, domestic chicks learn from gustatory cues to eat only foods that agree with them. Although taste is not unimportant to birds, they have far fewer taste buds than do mammals, which depend more heavily on this sense for choosing their diets. A chicken has only about 340 taste buds, a parrot 350, and a Mallard 375, far less than a rabbit, with 17,000, a rat with 1,265, a hamster with 725, or a human with about 9,000. On the chicken, Mallard, and other birds the taste buds are not mainly on the tongue but on the palate and lining of the mouth.

One might suppose that a bird, mostly covered with feathers, would be less sensitive to touch than are humans with bare skin. However, each feather is inserted into the bird's skin, which its movement would distort, thereby making the bird highly sensitive to tactile stimuli. Free sensitive nerve endings are well distributed over avian bodies, with touch-responsive corpuscles at the insertions of the contour feathers. These bodies are also present on palates and tongues, and are concentrated on the ends of woodpeckers' tongues and the bill-tips of birds, especially ducks and shorebirds. Even the scaly feet and toes of birds are touch sensitive. The skin of birds is also sensitive to pain and temperature. When too warm, they compress their plumage to increase its conductivity to heat; when cold, they fluff out their feathers to leave insulating air spaces among them.

The megapodes, or mound-birds, of Australia and islands of the south-

Mallee Fowls at incubation mound

western Pacific never sit on their eggs as other birds do but incubate them by volcanic or solar heat or that generated by decaying vegetable matter, singly or in combination. The Mallee-Fowl of semiarid regions of Australia maintains the internal temperature of its large mounds by skillful manipulation of solar heat and fermentation, as H. J. Frith demonstrated. To test the temperature, the birds thrust their bills into the mounds, apparently sensing the heat with either the interior of the bill or the tongue or both. The Brush-Turkey, which depends wholly on fermentation to warm its much larger mounds, pushes its whole bare red head into the decaying litter as R. S. Seymour and D. F. Bradford saw. So sensitive are the "thermometers" of these two species of megapodes that they can hold the temperature of their incubating eggs very close to the optimum of ninety-two degrees Fahrenheit (33° C). These birds, especially the Mallee-Fowl, appear to use intelligence and judgment to control their complex system of temperature regulation in all the vicissitudes of weather during their nine-month breeding season.

In addition to the five senses that they share with us, many birds have a sixth sense that we lack—sensitivity to magnetism, which has been investigated by H. H. Edwards and co-workers. Minute crystals of magnetite, an oxide of iron, are being found in the heads and necks of an increasing number of birds, but just how they perceive magnetism remains obscure. Exposing birds in their seasons of migration to artificially altered magnetic fields changes the direction they try to take, suggesting that sensitivity to terrestrial magnetism may be a principal means of orientation.

By watching living birds we learn how the information they receive through their six senses influences their behavior, and laboratory investigations reveal how sensory stimulation affects their brains and nervous systems; but we are far from understanding the relation of consciousness or thought to the brain. In discussing this problem, three related facts deserve attention. The first is that consciousness, often so acute in oneself, is never an observable phenomenon. We infer its presence in other creatures, human or different, from their resemblance to ourselves in structure and behavior.

The second, more surprising fact is that we lack direct awareness of where, within our own bodies, consciousness—our thoughts and feelings—is situated. All bright schoolchildren are quite sure that they think with their heads, and even a psychologist as sagacious as William James claimed to be immediately aware that his brain was the seat of his thought. Like most people these days, I accept the testimony of those who have investigated the anatomy of the nervous system and the effects of brain injuries on thought and sensations, that I do think with my brain, but I could never convince myself that I am directly aware of this fact. Those who claim immediate knowledge of the location of consciousness have apparently been conditioned to this belief by the widespread diffusion of the evidence from anatomy. Even the people who are so sure that the brain is the seat of consciousness cannot, I believe, tell us just where, within their capacious skulls, a certain thought or sensation is situated, surely not with the precision that they can locate a pinprick or mosquito bite on their skins.

Among the ancients, whose anatomy was at best rudimentary, Aristotle—no slovenly thinker—believed that reason resided in the chest; he regarded the brain as a sort of cooling device on top of the head. His pupil, Plato, envisaged a tripartite soul with the rational division in the head, the spirited or emotional part in the chest, and the appetitive soul that controls the organic functions lower in the body. Moreover, such everyday expressions as heartwarming, heartbreaking, heartfelt, heart-to-heart, to have a heart, and the like reveal clearly where for generations people have sponteneously located their sentiments.

The third salient fact is that, despite certain plausible hypotheses, we do not know how memories are stored in the brain and how they are reactivated. These three uncertainties should make us cautious about inferring the nature of an animal's psychic life from the characteristics of its brain.

One of the most admirable achievements of modern technology is miniaturization, especially valuable in an era when natural resources of many kinds are so lavishly exploited. Much smaller artifacts can now do more than older models could. Old-fashioned pocket watches, which needed twice-daily winding and could only tell the hour and its divisions, are replaced by smaller wrist watches, which are often automatic and can indicate the date as well as the time of day, and sometimes more. The newer radios and computers are not only more compact but also more efficient than earlier models, and so with many other inventions. Long before the advent of technology, nature began to miniaturize. An outstanding achievement is the hummingbird, a feathered jewel weighing only a few grams, with the most precise control of flight, able to hover motionless in the air as no other bird can, to economize energy by "noctivating" (reducing its temperature) on cool nights, to build charming nests, and (if it breeds in eastern North America) to fly nonstop across the six hundred miles (1,000 kilometers) of the Gulf of Mexico.

Another outstanding achievement of miniaturization is the avian brain. A huge whale could afford to carry much more cerebral tissue than it may need. We humans appear rarely to make full use of our brains; it has even been asserted that half the brain is a spare. Unlike swimming and ambulatory animals, flying birds cannot afford to carry excess weight; they are built for a maximum of strength and efficiency with a minimum of materials. It is probably to reduce weight slightly that birds' sexual organs, especially the testes, shrink after the breeding season. Not all birds live in the same way; differences in their lifestyles may be accompanied by differences in their brains. Birds that build elaborate nests may have certain cerebral tissues more developed than do those that make simple nests or none. Long-distance migrants probably have special neural connections. Birds with large song repertoires differ from those that sing little; neurons in the external striatum that control song vary in structure with the seasonal waxing and waning of singing.

When we contemplate all the capabilities of birds—their recognition of individuals, their memory, their skill in nest construction, the elaborate care that many take of their young, feats of navigation that, despite immense research, we still do not wholly understand—when we contemplate all that avian brains pack into such small compass, we must applaud them as among the most wonderful pieces of organized matter that nature has produced.

References: Bang and Cobb 1968; Campbell and Lack 1985 (articles on hearing and balance, nervous system, smell, taste, touch, and vision); Edwards et al. 1992; Frith 1962; Ridley 1952; Seymour and Bradford 1992; Skutch 1991; Stager 1964; Weidenreich 1948.

Chapter 15

❧

Homing
and Migration

A STUDY of birds' minds would be incomplete without looking briefly
at their marvelous feats of homing and migration. Taken from their nesting burrows on the Isle of Skokholm off the southwestern coast of Wales
and banded, two Manx Shearwaters were carried by air to Boston by
Rosario Mazzeo. The bird that survived the journey was released at the
water's edge. Twelve and a half days later, this shearwater returned to its
nesting burrow after crossing 3,200 miles (5,150 kilometers) of the North
Atlantic at an average speed of 250 miles (402 kilometers) per day. Manx
Shearwaters range far and wide over the Atlantic, but this bird had probably never been near the coast of New England.

In an earlier test of Manx Shearwaters' navigational ability, R. M.
Lockley sent two by air from Skokholm to Venice. Released there, one
reached home fourteen days and five hours later. If she took the most direct route to Skokholm, she covered 930 miles (1,497 kilometers), nearly
all overland. The improbability of such a journey by a marine bird that
does not normally ventures inland was diminished when three of twelve
birds released in the Swiss Alps returned to their nests on Skokholm in
ten to fifteen days.

Another tube-nosed swimmer, Leach's Petrel, nests on Kent Island,
near Grand Manan off the coast of New Brunswick, Canada. Susan Billings
took seven from their burrows, carried them by air to England, and released them in Sussex. They started westward, toward Canada, but were
blown southward. The great-circle distance from the point of release to
Kent Island is 2,980 miles (4,796 kilometers). The two fastest petrels returned home in 13.7 days, traveling at an average speed of at least 217 miles

(350 kilometers) per day. Another took about sixteen days to return, and the fourth about twenty four days. Three of the seven were not seen again.

Laysan Albatrosses lay their eggs on open ground on Sand Island, Midway Atoll, in the central North Pacific. Karl W. Kenyon and Dale W. Rice captured eighteen of these big birds and sent them by United States naval aircraft to points scattered in and around the ocean, where they were released on the water. Fourteen of these albatrosses returned to their nests from places from 1,315 to 4,120 statute miles (2,110 to 6,631 kilometers) away. The longest of their homeward journeys, from Luzon in the Philippine Islands, took 32.1 days, at an average speed of 128 miles (206 kilometers) per day. The next most distant point of release was on the coast of Washington, 3,200 miles (5,150 kilometers) from Sand Island, to which two of the four birds returned in 12.1 and 10.1 days, at average speeds of, respectively, 264 and 317 miles (425 and 510 kilometers) per day.

All these marine birds captured at their nests were experienced travelers, for they had returned, probably repeatedly, from wide expanses of the oceans where they foraged to the islets where they nested. Domestic breeds of the Rock Dove do not normally travel far, but by careful selection and training they can learn to return to their lofts from distant points. In pigeon races the best of them are occasionally released as much as 1,000 miles (1,610 kilometers) away, but only about one in twenty returns home.

Bank Swallows in Wisconsin demonstrated no comparable ability to return to their nests. When Theodore Sargent released these swallows individually at distances up to fifty miles (80 kilometers) from home, over 80 percent returned, but less than 40 percent returned from points fifty to one hundred miles (80 to 160 kilometers) distant. They appeared to find their way by searching randomly until they found familiar landmarks. Purple Martins, taken from nests in Michigan by William Southern, returned from distances up to 426 miles (686 kilometers). One female returned to her nestlings after flying 234 miles (377 kilometers) in 8.6 hours, mostly by night, which was unexpected in a diurnal migrant. Of twelve juveniles, released at distances up to 250 miles (402 kilometers), only two returned, from points in sight of the colony. Whether they failed to come home because their motivation or their ability to find their way was less than that of adults was not determined. In Europe, a Barn Swallow completed a homing flight of 1,150 miles (1,850 kilometers). All these members of the swallow family migrate for much greater distances than these artificially transported birds covered.

The procedure of removing marked birds from their nests, releasing them at precisely known points where they may never have been before, and carefully timing their return yields information not otherwise obtainable about the distances flown and the speeds of individuals. The perfor-

Manx Shearwater

mances of unmarked birds migrating between widely separated breeding and wintering areas cannot be measured with such exactness but are no less impressive. One of the greatest of feathered travelers is the Arctic Tern, the journeys of whom have been traced by F. C. Lincoln. After nesting from northwestern Greenland, the Arctic Archipelago of Canada, and southward to the coast of New England, many of these terns fly across the Atlantic to the west coast of Africa, thence southward to Antarctic waters. Their extreme summer and winter homes are 11,000 miles (17,703 kilometers) apart. After a circuitous trip of about 25,000 miles (40,234 kilometers), many return to their nest sites of the preceding year. No other bird enjoys so many hours of daylight.

From western Alaska where they nest, Bristle-thighed Curlews cross the Pacific to winter in Hawaii and Polynesia, including islands over 6,000 miles (9,656 kilometers) away, with minimal sea crossings of about 2,000 miles (3,220 kilometers). After scattering widely over the North and South Atlantic, Greater Shearwaters congregate in millions to nest on the Tristan da Cunha Islands, a small archipelago spread over only thirty miles (50 kilometers) of ocean and 1,500 miles (2414 kilometers) from the nearest land.

From the northern United States and southern Canada, Bobolinks migrate to southern Brazil, Bolivia, and northern Argentina, on the way crossing 500 miles (805 kilometers) of islandless water from Jamaica to northern South America. The extreme points of their breeding and winter ranges are about 7,000 miles (11,265 kilometers) apart. The eastern form of the Lesser Golden-Plover winters in the same region of South America but nests much farther toward the Pole, in Alaska, Canada, and the arctic islands. In the fall, adults fly eastward through Canada to the maritime provinces and New England, thence nonstop over the western Atlantic to South America. To return, they choose a more inland and westerly route through South and Central America and Mexico, then through the interior of the United States to Canada and beyond. In these journeys they describe a great ellipse with a major axis of about 8,000 miles and a minor axis of 2,000 miles (12,875 and 3,219 kilometers). The western race of the Lesser Golden-Plover breeds around the Bering Sea and migrates in the fall to Hawaii and Polynesia, in long overwater flights comparable to those of the Bristle-thighed Curlew.

Every year, to escape the harshness and scarcity of approaching winter, billions of birds from all over the north temperate zone and the Arctic fly thousands of miles over fertile lands, inhospitable deserts, and wide oceans to the tropics of America, Africa, Asia, Indonesia, Australia, even beyond in the south temperate zone, and, for some water birds, to cold Antarctic seas. For many of these birds, especially in the New World, migration is essentially homing, in two senses. In the first sense, southward migration is the return to an ancestral home. The hosts of hummingbirds, wood warblers, tanagers, orioles, vireos, and others that breed throughout temperate North America and winter farther south belong to families that originated in the American tropics, principally South America. As the last continental glacier receded, some ten thousand years ago, they extended their breeding ranges northward, attracted by abundant food in the newly ice-free lands, but each autumn they returned southward to avoid the cold, snowy gales that blew from the receding ice. Gradually, century by century, the pioneers helped to repopulate the vast areas that the glaciers had scraped free of life, until some reached high latitudes for nesting. All these summer residents are tropical birds that go north to breed, not, as was once thought, northern birds that fly south to escape winter's chill, and each autumn they return to the cradle of their race, where many of them spend much more time than they do where they nest.

Some of these migrants travel overland, probably retracing the route by which their ancestors extended their range northward. Others, from hummingbirds to wood warblers and tanagers, have shortened their journeys by flying nonstop across the 600-mile (1,000-kilometer) expanse of the Gulf of Mexico; while on their fall migration tiny Blackpoll Warblers, aided by favorable winds, fly over the Atlantic and Caribbean from north-

Black-and-White Warblers

eastern North America to northeastern South America, fueling this amazing continuous traverse of about 3,000 miles (4,828 kilometers) with fat deposited beneath their skin before they start. On their return to their breeding grounds in the spring, they choose a less strenuous route through the West Indies, Bahamas, and eastern United States.

Birds that have extended their ranges in more recent times exhibit the same tendency to retrace the routes their forerunners took. Bobolinks and Blackpoll Warblers, who from eastern North America colonized the far west, fly eastward before they turn southward, not without a tendency to shorten their route by a more direct southerly course. Arctic Warblers, members of an Old World family that through Siberia reached, and nest in, western Alaska, return through Asia to warmer parts of the Old World, parting company with Yellow Warblers, their neighbors in Alaska, who winter in tropical America with other wood warblers. Other Old World species that have invaded boreal America, the Bluethroat and Northern Wheatear, also migrate back to the eastern hemisphere.

In the southern hemisphere, where much less land lies at high latitudes than in the northern, migrations tend to be shorter and less spectacular. As the austral winter approaches, the relatively few migratory birds that breed in Tierra del Fuego, southern Chile, Patagonia, and central Argentina can travel overland by easy stages, straight northward to warm regions. In the Old World, where billions of birds that breed in Europe and western and central Asia winter in Africa south of the Sahara, the routes of many migrants are less direct, and more hazardous, than in the New World, including the traverse of vast, parched deserts as well as broad expanses of water. Their problems have been surveyed by P. Berthold and A. J. Helbig, with other authors in a special issue of *Ibis*.

First-time migrants and older ones face different problems and exhibit different degrees of precision. Young birds who have not yet claimed breeding territories in the north, and certainly do not know exactly where they will spend the winter in the south, home mainly in the sense of returning to the land of their ancestors. Many older birds who have established both summer and winter territories fly back and forth between them until they die, yearly accomplishing feats of long-distance homing unequaled by any domestic pigeon. Young birds on their first southward or northward migration need only to fly in the right direction, with perhaps changes in bearing along the way, and to know when to stop, lest they go beyond the ranges of their species. Their "bearing and distance" navigation is innate. To what degree they are guided on their maiden flights by the call notes of the many species of small migrants that fill the nocturnal skies along the flyways is uncertain.

More prone to stray than are experienced migrants, many juveniles fly too near the Atlantic or Pacific coasts of the United States and wander or are blown out to sea, where large numbers may be lost. However, many also reorient and reach shore, where during the fall migration disproportionately large numbers of juveniles have been found by B. G. Murray, Jr., and other bird-banders. When they return to their breeding ranges in the spring, some young seek the vicinity of their birthplaces, where more Prairie Warblers than could be attributed to chance were found by Val Nolan, Jr.; but the few yearlings seen there contrasted with the many older birds that reclaimed their former territories year after year. Young birds appear to scatter widely over the range of each species. To reach a goal as wide as Mexico, Central America, or northern South America in autumn, or eastern North America in the spring, hardly requires very precise navigation. The tendency of the young to scatter promotes outbreeding and range extension.

In contrast to the young, experienced birds navigate with amazing precision. How can a bird without a single instrument find its objective as well as an airplane pilot with a whole panel of them? Their ability to find a known acre or two of land three thousand miles away depends in part on learning. When we remember that domestic pigeons are trained to find their way home from increasingly distant points of release, it is reasonable to believe that long-distance migrants learn to find precise localities, the difference being that they are self-taught.

Not the least of the remarkable aspects of birds' migrations is their ability to correct their courses if they stray. Blown off course by adverse winds, or simply missing their way, as sometimes happens to inexperienced juveniles, many nocturnal migrants find themselves over unfamiliar seas as night ends. Radar, for nearly four decades widely used to study migration, has revealed that, after descending lower after midnight, nocturnal travelers ascend around dawn. From high above the sea, sometimes beyond sight of land, many of these birds take a new direction to reach shore. Thus, observing by radar in the Shetland Islands the behavior of Scandinavian thrushes migrating on autumn nights across the northeastern Atlantic Ocean from Norway, M. T. Myres, learned that after their dawn ascent, they veered southward to reach land.

On the coasts of Nova Scotia, John Richardson's radars revealed that autumn migrants, flying southwestward over the Atlantic from Newfoundland, often turned west-northwest or north-northeast toward the nearest land. Blown offshore by northwest winds in the fall, passerine migrants find refuge on Nantucket, Block Island, and others off New England's southern coast. After resting there, they fly northward at dawn or in the early morning to reach the mainland, as studies by J. Baird and I. C. T. Nisbet, K. P. Able, and others have revealed. Wood warblers, who reached the coast of Louisiana after flying northward across the Gulf of Mexico while winds blew from the east in spring, compensated for their westward displacement by choosing a northeasterly course when they resumed their migration, as F. R. Moore has shown.

Reorientation appears to require abilities and sources of information not ordinarily needed by juveniles, who maintain a genetically determined direction from the breeding to the wintering ranges of their species, nor by adults retracing familiar routes between their nesting and wintering territories. As a rule, the reorienting bird does not take the most direct course to its ultimate goal from the point where it finds itself when it becomes aware that it has strayed, but it tries to regain, as quickly as possible, an established flyway, even if it must fly northward at a season when its usual course is southward. The young bird's innate tendency to fly toward its migratory goal would often prove inadequate if not supplemented by the ability to reorient when blown off course. Since no innate directive could foretell all the vagaries of the winds, it is evident that to reach its destination, even a broad landmass, birds need minds capable of guiding them in diverse predicaments.

A retentive memory of details of the landscape visible from high in the air helps birds to reach a familiar spot within it. For greater distances, "leading lines," such as long mountain ranges, great rivers, or the seacoast may help guide diurnal migrants to their destinations, and also nocturnal migrants on nights not too heavily overcast. But these geographical features would help many birds for only part of their long way, and beyond sight of land over the oceans they would be unavailable. To direct their diverse courses, birds need means of orientation more uniformly distributed over the planet, one of which is the Sun's position in the sky. With the help of their internal clocks, birds can compensate for its hourly changes in altitude and horizontal direction, as, in the north temperate zone in autumn, to fly southward with the Sun on the left at sunrise, with it ahead at midday, on the right at sunset, and with finer corrections for intermediate hours. As earlier noted, changing homing pigeons' internal clocks by exposing them to different schedules of artificial lighting causes them to take directions that are correct by their altered time but lead them far astray.

Except kingbirds, swallows, and a few other aerial flycatchers that eat as they travel, most small birds migrate by night, avoiding the many diurnal raptors and dehydration in strong sunshine, and in the daytime interrupt their journeys to forage and rest. These birds use the stars for orientation, as has been demonstrated by many experiments. If confined in a cage when they should be traveling, they exhibit migratory restlessness, often designated by the German *Zugunruhe,* trying to escape in the direction they would take if free—in northern birds to the south in fall, to the north in spring. In a planetarium with normal setting, as under a starry sky, these are the directions that most of the captives choose. If the setting of the planetarium is changed, as by showing the September constellations in March or the March constellations in September, the birds' *Zugunruhe* is also reversed; they try to fly northward in autumn and southward in the spring. It appears that they recognize and are guided by the Pole Star with its constant position amid the revolving constellations. When, in a planetarium, the heaven is made to circle around some other star as center, the birds orient by this star as though it were Polaris.

Stephen Emlen demonstrated that young Indigo Buntings lack an innate map of the nocturnal sky but learn to recognize the Pole Star (or its substitute in a planetarium) by observation, probably of the constellations' changing positions rather than by their slow progress. One wonders whether, as the date of their migration approaches, birds who usually roost amid foliage perch for a while in the open, studying the heavens, which they probably did ages before Chaldean and Babylonian astrologers named the constellations. We do not know how nocturnal migrants guide themselves when they fly so far south that Polaris sinks below the horizon.

Probably at latitudes where the Pole Star is no longer visible, or when it is obscured by clouds, migrants set their courses by their sensitivity to Earth's magnetic field, the most constant and reliable indicator of direction, as the compass needle is for us. The minute crystals of magnetite in their heads may act as little lodestones, pressing against highly sensitive nerve endings. Tiny magnets attached to their heads disorient pigeons flying under an overcast sky but brass bars of equal weight do not. Mounted on pigeons' heads, electric coils that change the direction of the magnetic field alter their courses. The direction in which enclosed migrants try to fly can be changed by shifting their magnetic field, much as it is changed by showing them different star patterns in a planetarium. In one of the more recent tests of the effect of magnetism on bird migration, Verner Bingman demonstrated that, in a vertical magnetic field that provided no directional information, confined European Robins tried to fly in the direction determined by the horizontal field to which they had been previously exposed. To be so well equipped with directional cues that if one fails they can use another must save many migratory birds from going astray.

For precise long-distance navigation, as practiced by birds who each year commute between pinpoints on the map thousands of miles apart, a reliable direction-giver is not enough; a chart or its equivalent is indispensable. Navigators need to know their exact position, in terms of latitude and longitude or otherwise, that of the destination, and the direction of the second from the first. Over short distances, random searching for familiar landmarks may be adequate, but over the vast areas traversed by migrating birds it would be far too expensive of time and energy. Despite much research and some elaborate, unconfirmed hypotheses, this great mystery of migration remains unsolved; we still do not know what birds use for a chart.

Whether they return to a well-hidden nest through the mazes of a forest where we are so readily lost or to a far distant territory, birds reveal a rapport with Earth and all its subtle signs and emanations that most contemporary humans lack, although in past ages our ancestors may have been more sensitive to them. Beginning with the innate capacity for bearing and distance travel, birds learn the finer art of precision navigation, in which they are aided by acute sensitivity to a diversity of direction-givers, by close attention to topographical features, and by the ability to remember features from year to year. Could they find a tiny islet in the midst of a vast ocean, or a familiar acre or two amid wide forests or farmlands, without finely organized brains, keen senses, and awareness of what they do?

References: Able 1977; Baird and Nisbet 1960; Berthold and Helbig 1992; Billings 1968; Bingman 1987; Campbell and Lack 1985 (articles homing, migration, navigation, and pigeon); Emlen 1967, 1970; Kenyon and Rice 1958; Lincoln 1950; Lockley 1942; Mazzeo 1953; Moore 1990; Murray 1966; Myres 1964; Nolan 1978; Richardson 1978; Sargent 1962; Skutch 1991; Southern 1968.

Chapter 16

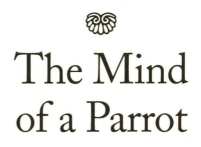

The Mind of a Parrot

BETTER COMMUNICATION, within and between species, would be a blessing to the living world. A universal language would facilitate learning, help to reduce tribalism, and improve international relations. A farmer who could communicate with his domestic animals might persuade them to respect his corn and other crops, and to cooperate instead of resisting when he tries to cure their ailments, for the benefit of all concerned. We might even cultivate better relations with the free animals around us if we could understand one another. Communication could not eliminate all conflicts of interest, but with understanding and goodwill it should mitigate them. To naturalists and psychologists trying to fathom the minds of animals, free communication would be a boon.

The great apes are widely held to be the most intelligent of nonhuman terrestrial animals, but their inability to pronounce human words makes it necessary for researchers laboriously to learn the sign language of the deaf, then patiently teach it to the chimpanzee, gorilla, or orangutan with whom they wish to communicate. The many mimetic birds who can imitate the most diverse sounds and, moreover, are easier to procure and to keep, should be more promising subjects for studies of animals' mental capacities; but to pronounce words or repeat sentences is not equivalent to understanding their meaning. In the mid-twentieth century, Konrad Lorenz could write that his tame Raven, Roah, "is, so far as I know, the only animal that has ever spoken a human word, to a man, in its right context—even if it was only a very ordinary call-note."

Long before this, parrots and other birds had correctly pronounced human words, but without much indication that they understood what they said. By the 1940s, researchers well trained in the methods of the psychology laboratory had taught mimetic birds to repeat human words, but the birds did not learn their meanings. This failure was not due to the birds' dullness so much as to the inappropriate methods used to instruct them by investigators unfamiliar with the way free birds learn their repertoires. It remained for Irene Pepperberg, who after receiving her doctorate in chemical physics switched to the equally arcane subject of animal psychology, to perfect a method for teaching birds to use human speech with understanding. Thereby, she opened the way to reciprocal communication between birds and humans, with impressive revelations of their mental ability.

The subject, or might we say the collaborator for a dozen years, of Pepperberg's researches is a long-lived African Gray Parrot that she chose at random in a pet shop when he was thirteen months old. Born in captivity but not hand-raised, Alex is not as closely bonded to humans as rearing by them might have made him. When bored by his lessons, he may refuse to answer a question, shout "No!" and turn his back to his instructors. He may interrupt a session by demanding food, a toy, or to be taken somewhere else. When left with a new student trainer, he may request, one by one, dozens of items from his array of toys and foods, while the student scurries around to fetch them. Alex appears to be testing his new associate's knowledge of the words he has learned.

The method chiefly employed to demonstrate to Alex the meanings of words and phrases has been variously called social modeling or the model-rival technique. It involves three-way interactions among two competent human speakers and the avian student. One person is the trainer, while the other acts as both the model for the bird's responses and the rival for the trainer's attention. Holding one or more objects in front of both the bird and his rival, the trainer asks, "What's here?" "What color?" "How many?" or similar questions, and rewards the correct answer with praise and the object(s). This procedure teaches the parrot the names, attributes, or quantities of things, or on other occasions it demonstrates actions and the names for them.

When the rival purposely imitates the bird's incorrect answer or unclear pronunciation, the trainer scolds him and removes the object from sight, showing the bird the consequences of his mistakes. The model-rival is requested to try again or to speak more clearly, thereby helping the parrot to correct his error. Then the act is repeated with trainer and rival reversing their roles, the parrot sometimes participating. This change of roles prevents the bird from restricting his responses to the particular person who asked the question, and helps him learn both parts of the interaction. Finally, the bird is given an intrinsic reward—the item that he correctly named or described—instead of something unrelated to his task,

African Gray Parrot

as when in laboratory tests the subject receives food for pecking a button or moving the lever that releases hidden food. Such unrelated rewards are less effective in promoting learning. In these tests, every precaution was taken to ensure that the parrot was independently performing the task assigned to him, instead of acting upon some extraneous cue inadvertently provided by his trainer, in the manner of Clever Hans, the horse that appeared able to solve mathematical problems.

The first step in Alex's training was to teach him the English names of various objects and their attributes. He was an apt pupil, learning the labels for forty objects and materials, including paper, key, wood, cork, nut, banana, carrot, chain, water, showah (shower), and thirty others. He could name seven colors, from rose (red) to purple. He could identify five shapes by the number of their angles, as two-, three-, four-, five-, or six-cornered objects. Moreover, he could combine all his vocal labels to identify correctly, request, refuse, and categorize about one hundred diverse things, including some differing from the examples with which he had been trained, or exhibiting novel combinations of attributes, as on boxes of different colors or variously shaped pieces of woolen fabrics. His accuracy averaged about 80 percent and might have been higher if he had not occasionally been recalcitrant. With this vocabulary, Alex was prepared to demonstrate higher levels of mental ability.

The model-rival procedure was also used to show Alex that he could not obtain something by simply naming it; he must indicate his desire by the word *want*. He learned to say, "I want x," "Want x," "Wanna x," or

African Gray Parrot

"Wanna go y" where x and y are appropriate labels for objects or localities. He wanted the key that he won by correctly naming it for scratching the area around his bill (tool using), or he wanted a cork for chewing and his colorful toys for playing. If handed something that he did not want, he might reject it with "No!" and repeat his request.

Alex could name seven colors when he saw them on familiar objects, but could he form the concept "greenness," thereby recognizing what green grass and a green key have in common? Was he aware that all these colors are included in a class or category "color"? He could name five shapes, but could he unite them in the category "shape"? And could he mentally separate these two concepts? To answer these questions, Alex was shown familiar and unfamiliar objects with diverse combinations of color and shape, and asked, "What color?" or "What shape?" To respond correctly, Alex had first to notice the different recognizable attributes of the things presented to him. He had also to understand to which of these categories he was requested to attend. Next, he must correctly name the attribute in the category that the object exhibited. He must also identify the material that bore the attribute, so that he might reply, "Green wood" or "Two-corner hide." In these tests he responded with an accuracy of over 80 percent, demonstrating that he could categorize, distinguish different categories, recognize instances of each, and combine them correctly, as when he said, "Five-corner wood."

The parrot could tell which objects were alike or different in color, shape, or material. When there was no difference, he would declare, "None," revealing that he had some concept of absence. When confronted with an array of things differing in size, shape, color, and material, he could answer the questions, "Which is larger?" or "Which is smaller?" by naming the color or material of the bigger or smaller object; and when there was no difference in size, he would reply, "None." He answered these questions with an overall accuracy of 78.7 percent, demonstrating that he had grasped the concept of size. As told in chapter 6, Alex could count up to six, apparently he was not taught beyond this number.

Studies of primates' intelligence have occupied more scientists for more years than has Pepperberg's path-breaking work with Alex, raising their investigations to higher levels of complexity; but as far as his mental range has been tested, the African Gray Parrot does not fall below the great apes. Alex should become as famous as Washoe, the chimpanzee, and Koko, the gorilla, both with much bigger brains in much bigger bodies. Kyaaro and Alo, two Gray Parrots that Pepperberg has raised by hand, should become more cooperative than Alex, who lacked this initial advantage. With them, she and her students hope to reveal mental aptitudes higher than Alex has so far demonstrated.

Pepperberg doubts that training in interspecific communication increases the intelligence of a bird, such as its ability to acquire a concept that otherwise it would be incapable of learning. But language, by facilitating communication between investigator and subject, makes the bird's mind more accessible to study, and permits the assignment of tasks that reveal its mental capacity. Without human language, pigeons could form open-ended categories (chapter 11), and a Great Tit could count up to eight (chapter 6). We have no reason to believe that parrots are exceptionally intelligent birds. Others are equally good or better mimics, build more elaborate nests, perform feats of navigation not yet demonstrated by a parrot, and live in societies more sophisticated than any yet reported of parrots. Other species may equal or surpass Alex in native intelligence and ability to learn, but without the communication that human language facilitates, this might be difficult to demonstrate. Their shorter lives might not permit the many years of instruction that Alex has received. Long-lived, readily domesticated parrots are exceptionally favorable subjects for the study of avian intelligence, which Pepperberg has pursued with outstanding dedication, patience, and insight. Laboratory research such as hers and that of other students of birds' psychology, complemented by sympathetic, understanding study of free birds in their native habitats, should finally convince us that birds have much more intelligence than has commonly been recognized.

New World parrots may be no less intelligent and capable of learning from human teachers than is the African Gray Parrot. In *Birds of Brazil*, Helmut Sick told of a Turquoise-fronted Parrot who was taught to distinguish seven shapes and numbers of dots in order to reach a hidden reward. Even an individual more than forty years old could learn to do this, although not as readily as younger parrots of the same species. This *Amazona* is reputed to be the best talker among Brazilian parrots, especially when its training begins at an early age. Relative to the weight of its body, the big-headed Blue-and-Yellow Macaw, widespread in South America, has the largest brain known among birds, exceeding that of members of the crow family. Macaws can be trained to ride bicycles, respond to questions, and select a red object among several of different colors.

References: Kaufman 1991; Lorenz 1952; Pepperberg 1985, 1990a, 1990b, 1990c; Pepperberg and Brezinsky 1991; Sick 1993.

Chapter 17

❦

Summary
and Conclusions

THE PRECEDING chapters offer many reasons for believing that birds' mental capacities have been grossly underestimated. Let us bring the most important of them together for a final survey:

Birds have a remarkable capacity to recognize individuals, not only other birds who to us appear indistinguishable but also humans. This ability is indispensable for their advanced social lives. They also have retentive memory of individuals, places, songs, and events, and they give indications that they anticipate the future.

Only humans take more elaborate care of their young than birds do. Cooperative breeders, of which increasing numbers are found as more tropical and subtropical species are studied, exhibit family unity equaled only in the most united families of humans and a few other mammals, including recognition of individuals and their relationships, and avoidance of incest.

In courtship and rearing young, birds give indications of emotions and affection that imply consciousness. In play, birds reveal their capacity for enjoyment, which is one of the strongest indications that they are conscious. The mental conflicts arising in their lives probably intensify their consciousness, as such conflicts do in us.

Birds can be taught to count up to eight, by tapping out the numbers with their bills or naming them in human language. They have a good sense of time, responsive to treatment that shifts their internal clocks. Tool using, known in a few birds, was apparently originated by inventive minds. In some cases it has become innate, probably by genetic assimilation.

The beautiful and often profuse plumage of many male birds that court in assemblies would be suppressed rather than favored by natural selection because it is expensive of materials, makes its wearers more conspicuous, impedes movement, and makes them unfit them for nest attendance. Female choice is the most convincing explanation of extravagant nuptial adornments. This, and the ornamentation of their pavilions by bowerbirds, point strongly to an aesthetic sense, as do the melodious songs of many species. They appear to enjoy singing.

Many birds resort to dissimulation to lure potential predators from their nests or fledged young, mainly by injury feigning. Whether they are aware that they are using deception is not known; but numerous observations leave no doubt that they are in full control of their movements and aware of what they do. As a rule, they perform the act only when they have a proper stage where they will not be entangled in vegetation, keeping just far enough ahead of the enemy to lure it away from the nest with the prospect of a meal, without incurring the danger of being caught. Scarcely anything else that birds do demands such cool calculation and quick wits.

Birds further reveal intelligence by behaving in ways favorable to themselves or to associates but not programmed in their genes. Because they are so well equipped with innate patterns of behavior adequate for their needs in natural environments, these innovations are most likely to occur, and to be recorded, in interactions with people and our constructions. In laboratories, pigeons and parrots have demonstrated their ability to form recepts, or unnamed categories, which are probably not foreign to the minds of free-living birds. Admittedly in some situations, again usually those created by people, birds act in ways that appear stupid. But such actions can often be attributed to behavior that is adaptive in natural circumstances. Occasional stupidity, moreover, does not negate intelligence either in birds or in humans.

Birds demonstrate their freedom from strict control by their genes in several ways. Feeding young or otherwise helping at nests of unrelated pairs, of the same or different species, is the exercise of an innate behavior—placing food into gaping mouths—in contexts not programmed in the genes. Puzzling to biologists, altruism presents no more difficulty to evolutionary theory than do play and artistic creation, which likewise use excess energy in satisfying activities. By varying the order of phrases in their songs, birds also demonstrate their freedom.

The brains of birds compare favorably in size with those of mammals of the same bodily size but are of somewhat different construction. They are superb examples of miniaturization, with intricate functions packed tightly together to save weight when flying. In addition to the five senses that they share with us, at least some birds are sensitive to Earth's magnetic field. In homing and migration, birds use a number of directional cues, including the Sun's position in the sky, the stars, terrestrial magnetism, major geographic features, probably odors in certain cases, and memory of local topography.

Irene Pepperberg has demonstrated that the mentality of an African Gray Parrot is in some respects of the same grade as that of chimpanzees, gorillas, and other great apes, held to be the most intelligent of nonhuman animals. She has shown how greatly the revelation of birds' innate capabilities depends upon appropriate methods of communicating with them.

Not every species exhibits all the abilities in this list. Advanced cooperative breeders do not migrate, and migratory birds do not live in cooperative groups. Only a few species have been watched using tools, and many do not sing melodiously; but together these capacities reveal the range and versatility of the avian brain and mind.

Although impressive, this incomplete list of birds' accomplishments is not adequate to prove to the skeptical scientist or philosopher that birds are conscious. The scientist might demand that their consciousness be made public and observable, either directly by human senses or by means of instruments; but despite the admirable ingenuity of modern instrument makers in devising the most sensitive apparatus, none to demonstrate the presence of consciousness is now available, or seems likely to be in the foreseeable future. The philosopher, if a thorough skeptic, would insist on being certain of nothing except one's own existence as a thinking being, of which one's very doubts of the existence of anything else provide irrefragable testimony, for doubting is a mental activity of an existing being. But a consistently skeptical philosopher or scientist would not deny that birds, or humans other than themselves, are conscious. Both would defer judgment.

We intuitively ascribe consciousness to other people and to at least some of the animals around us, with greater conviction the more they resemble us in structure and behavior, the more responsive they are to our attempts to be friendly. Birds and at least the more familiar mammals fall into this category. To all the foregoing reasons for believing that birds are conscious, evolutionary considerations add another of great weight. Birds' minds evolved to adjust innate behavior to the variables of heterogeneous environments. Since conscious awareness should increase their effectiveness, it is reasonable to believe that evolution would sharpen it. Indeed, if birds were wholly insentient, they could not properly be said to have minds, but only brains filled with neurons and synapses that connect them into an intricate network, which directs their courses through a complex and perilous world, much as an automatic pilot steers an airplane, but more subtly.

Why should we be so concerned about the minds of birds, their capacity for enjoyment and suffering? Many ornithologists appear to have little interest in the psychic life of the birds whose activities they minutely describe and try to reconcile with evolutionary theory, or try to explain in the manner of the behaviorists. Scientists and bird-watchers are moved by the appearance and melodies of birds and concerned for the conservation of particular populations and species for the sake of ecosystem integrity. But there may be more profound reasons for us to focus on the minds of birds. We are sometimes depressed by the suspicion that humans are the only intelligent, perhaps the only truly conscious beings, in a terrifyingly vast, purposeless universe. At great expense, we build huge disk antennas directed into outer space, with only the slightest probability of picking up intelligible messages from technically advanced inhabitants of planets light-years distant. We attach symbols to spaceships voyaging beyond our solar system, with faint hope that they will be found and interpreted by understanding beings. If they exist, the intelligent inhabitants of a planet that may be very different from ours probably do not exemplify the vertebrate type of organization and may be inconceivably alien to ourselves.

To alleviate our loneliness, we might devote more attention to accessible creatures who do have strong affinities with ourselves. The more profoundly and sympathetically we study them, the stronger grows our intuition that they are conscious, until it becomes almost a certainty, and the less remote from us they appear. A first consequence of our growing rapport with them is their inclusion in the sphere of our ethical concern; we treat them as creatures that enjoy and suffer, not as insentient objects.

Moreover, our worldview brightens as our sympathies expand. We recognize that as harmonization, the cosmic process, builds the materials of the universe into patterns of increasing amplitude, complexity, and coherence, the value of existence increases. Evolution, a phase of the cosmic process that is often excessively harsh, appears more benign when we recognize that, while diversifying the living world and adapting organisms to their environments, it makes many of them capable of enjoying their lives, and that this beneficent movement began long before we appeared on Earth. Trust in the beneficence of the cosmic process should help dispel the alienation and gloom that oppresses the human spirit in a world afflicted with countless ills, largely caused by our own lack of moderation in reproduction and in consumption of natural resources. It should also encourage us to intensify our struggles to save an overpopulated, overexploited planet from impending disaster, and especially to protect the birds, whose minds are among nature's greatest wonders.

Bibliography

Able, K. P. 1977. "The Orientation of Passerine Nocturnal Migrants Following Offshore Drift." *Auk* 94:320–30.

Alexander, J. R., and W. T. Keeton. 1974. "Clock-shifting Effect on Initial Orientation of Pigeons." *Auk* 91:370–74.

Armstrong, E. A. 1940. *Birds of the Grey Wind.* London: Oxford University Press.

———. 1942. *Bird Display: An Introduction to the Study of Bird Psychology.* Cambridge: University Press.

Ashmole, N. P., and H. Tovar S. 1968. "Prolonged Parental Care in Royal Terns and Other Birds." *Auk* 85:90–100.

Baird, J., and I. C. T. Nisbet. 1960. "Northward Fall Migration on the Atlantic Coast and Its Relation to Offshore Drift." *Auk* 77:119–49.

Balda, R. P. 1965. "Loggerhead Shrike Kills Mourning Dove," *Condor* 67:359.

Balda, R. P., and W. Wiltschko. 1991. "Caching and Recovery in Scrub Jays: Transfer of Sun-compass Directions from Shaded to Sunny Areas." *Condor* 93:1020–23.

Baldwin, M. 1974. "Studies of the Apostle Bird at Inverell. Part I: General Behaviour." *Sunbird* 5:77–88.

Bang, B. G., and S. Cobb. 1968. "The Size of the Olfactory Bulb in 108 Species of Birds." *Auk* 85:55–61.

Beck, B. B. 1986. "Tools and Intelligence." In *Animal Intelligence,* R. J. Hoage and L. Goldman, eds. Washington, D.C.: Smithsonian Institution Press, 135–47.

Bent, A. C. 1946. *Life Histories of North American Jays, Crows, and Titmice.* U.S. Natl. Mus. Bull. no. 191.

Bent, A. C., and Collaborators. 1968. *Life Histories of North American Cardinals, Grosbeaks, Buntings, Towhees, Finches, Sparrows, and Allies.* U.S. Natl. Mus. Bull. no. 237, part 1.

Berthold, P., and A. J. Helbig. 1992. "The Genetics of Bird Migration: Stimulus, Timing, and Direction." *Ibis* 134, suppl. 1:35–40.

Billings, S. M. 1968. "Homing in Leach's Petrels." *Auk* 85:36–43.

Bingman, V. P. 1987. "Earth's Magnetism and the Nocturnal Orientation of European Robins." *Auk* 104:523–25.

Bluhm, C. K. 1985. "Mate Preferences and Mating Patterns of Canvasbacks *(Aythya valisineria)."* In *Avian Monogamy,* P. A. Gowaty and D. W. Mock, eds. Ornithol. Monogr. no. 37:45–56. Washington, D.C.: American Ornithologists' Union.

Brackbill, H. 1961. "Shadow Boxing by Brown-headed Cowbirds." *Auk* 78:98–99.

Brown, J. L. 1987. *Helping and Communal Breeding in Birds: Ecology and Evolution.* Princeton, N. J.: Princeton University Press.

Brown, L. H. 1958. "The Breeding Biology of the Greater Flamingo *Phoenicopterus ruber* at Lake Elmenteita, Kenya Colony." *Ibis* 100:388–420.

Brown, L. H., and E. K. Urban. 1969. "The Breeding Biology of the Great White Pelican *Pelecanus onocrotalus roseus* at Lake Shala, Ethiopia." *Ibis* 111:199–237.

Brunton, D. H. "Fatal Antipredator Behavior of a Killdeer." *Wilson Bull.* 98:605–607.

Campbell, B., and E. Lack, eds. 1985. *A Dictionary of Birds.* Calton, England: T. and A. D. Poyser.

Catchpole, C. K. 1980. "Sexual Selection and the Evolution of Complex Songs among European Warblers of the Genus *Acrocephalus." Behaviour* 74:149–65.

Cherrie, G. K. 1916. "A Contribution to the Ornithology of the Orinoco region." *Brooklyn Inst. Arts and Sci.; Sci. Bull.* 2:133a–374.

Chisholm, A. H. 1954. "The Use by Birds of 'Tools' or 'instruments.'" *Ibis* 96:380–83.

Collias, N. E. 1952. "The Development of Social Behavior in Birds." *Auk* 69:127–50.

Conover, M. R. 1985. "Foreign Objects in Bird Nests." *Auk* 102:696–700.

Cornford, F. M. 1937. *Plato's Cosmology:* The Timaeus *of Plato Translated with a Running Commentary.* London: Routledge and Kegan Paul.

Coulson, J. C. 1972. "The Significance of the Pair Bond in the Kittiwake." *Proc. 15th Interntl. Ornithol. Congr.,* 424–433.

Coulter, M. C. 1980. "Stones: An Important Incubation Stimulus for Gulls and Terns." *Auk* 97:898–99.

Counsilman, J. J. 1977. "A Comparison of Two Populations of the Grey-crowned Babbler," part 1. *Bird Behaviour* 1:43–82.

Darling, F. F. 1937. *A Herd of Red Deer.* London: Oxford University Press.

Darwin, C. 1871. *The Descent of Man and Selection in Relation to Sex.* New York: Modern Library reprint.

Davis, D. E. 1940. "Social Nesting Habits of the Smooth-billed Ani." *Auk* 57:179–218.

Davis, E. R. 1926. "Friendly Siskins." *Bird Lore* 28:381–88.

Davis, L. S. 1982. "Timing of Nest Relief and Its Effect on Breeding Success in Adélie Penguins." *Condor* 84:178–83.

———. 1988. "Coordination of Incubation Routines and Mate Choice in
Adélie Penguins. *(Pygoscelis adeliae)*." *Auk* 105:428–32.

Dickinson, J. C., Jr. 1969. "A String-pulling Tufted Titmouse." *Auk* 86:559.

Dowsett-Lemaire, F. 1979. "The Imitative Range of the Song of the Marsh
Warbler *Acrocephalus palustris*, with Special Reference to Imitations of
African Birds." *Ibis* 121:453–68.

Edwards, H. H., G. D. Schnell, R. L. Dubois, and V. H. Hutchison. 1992.
"Natural and Induced Remanent Magnetism in Birds." *Auk* 109:43–56.

Elliot, R. D. 1977. "Hanging Behavior in Common Ravens." *Auk* 94: 777–78.

Emlen, S. T. 1967. "Migratory Orientation in the Indigo Bunting, *Passerina
cyanea*." Part 1: Evidence for the Use of Celestial Cues. Part 2: Mecha-
nism of Celestial Orientation. *Auk* 84:309–42, 463–89.

———. 1970. "Celestial Rotation: Its Importance in the Development of
Migratory Orientation." *Science* 170:1198–1201.

Evans, R. M. 1970. "Parental Recognition and the 'Mew Call' in Black-
billed Gulls *(Larus bulleri)*." *Auk* 87:503–13.

Ficken, M. S. 1977. "Avian play." *Auk* 94:573–82.

Ficken, M. S. and R. W. Ficken. 1987. "Bill Sweeping Behavior of a Mexican Chickadee." *Condor* 89:901–902.

Fox, W. 1952. "Behavioral and evolutionary significance of the abnormal
growth of the beaks of birds." *Condor* 54:160–62.

Frazer, J. G. 1944. *The Golden Bough: A Study in Magic and Religion.* New
York: Macmillan.

Friedmann, H. 1934. The Instinctive Emotional Life of Birds. *Psychoanalytic
Review* 21:1–57.

Friedmann, H., and L. Kiff. 1985. "The Parasitic Cowbirds and Their
Hosts." *Proc. Western Foundation of Vertebrate Zoology* 2:225–302.

Frith, H. J. 1962, *The Mallee-Fowl: The Bird That Builds an Incubator*
Sydney: Angus and Robertson.

Gaston, A. J. 1977. "Social Behaviour within groups of Jungle Babblers
(Turdoides striatus)." *Anim. Behav.* 25:828–48.

Gayou, D. C. 1982. "Tool use by Green Jays." *Wilson Bull.* 94:593–94.

Gilliard, E. T. 1969. *Birds of Paradise and Bower Birds.* London: Weidenfeld
and Nicolson.

Goodwin, D. 1978. *Birds of Man's World.* Ithaca, N.Y.: Cornell University Press.

Grimes, L. G. 1976. "The Occurrence of Cooperative Breeding Behaviour
in African Birds." *Ostrich* 47:1–15.

Guhl, A. M., and L. L. Ortman. 1953. "Visual Patterns in the Recognition
of Individuals among Chickens." *Condor* 55:287–98.

Gullion, G. W. 1954. "The Reproductive cycle of the American Coot in California." *Auk* 71:366–412.

Hamilton, W. J., III. 1962. "Evidence Concerning the Function of Nocturnal Call Notes of Migratory Birds." *Condor* 64:390–401.

Hardy, J. W., T. A. Webber, and R. J. Raitt. 1981. "Communal Social Biology of the Southern San Blas Jay." *Bull. Florida State Mus., Biol. Sci.* 26:203–63.

Hartshorne, C. 1973. *Born to Sing: An Interpretation and World Survey of Bird Song.* Bloomington: Indiana University Press.

Higuchi, H. 1986. "Bait-fishing by the Green-backed Heronm *Ardeola striata* in Japan." *Ibis* 128:285–90.

————. 1988. "Individual Differences in Bait-fishing by the Greenbacked Heron *Ardeola striata* Associated with Territory Quality." *Ibis* 130:39–44.

Hoage, R. J., and L. Goldman, eds. 1986. *Animal Intelligence: Insights into the Animal Mind.* Washington, D.C.: Smithsonian Institution Press.

Howard, L. 1952. *Birds as Individuals.* London: Collins.

————. 1956. *Living with Birds.* London: Collins.

Hundley, M. H. 1963. "Notes on Methods of Feeding and the Use of Tools in the Geospizinae." *Auk* 80:372–73.

Immelmann, K. 1966. "Beobachtungen an Schwalbenstaren." *J. für Ornith.* 107:37–69.

Jaeger, E. C. 1951. "Pebble-dropping by House Sparrows." *Condor* 53:207.

Janes, S. W. 1976. "The Apparent Use of Rocks by a Raven in Nest Defense." *Condor* 78:409.

Johnstone, R. M., and L. S. Davis. 1990. "Incubation Routines and Foraging-trip Regulation in the Grey-faced Petrel *Pterodroma macroptera gouldi.*" *Ibis* 132:14–20.

Jones, T. B., and A. C. Kamil. 1973. "Tool-making and Tool-using in the Northern Blue Jay." *Science* 180:1076–77.

Kaufman, K. 1991. "The Subject is Alex." *Audubon* 93(5):52–58.

Kemp, A. C., and M. I. Kemp. 1980. "The Biology of the Southern Ground Hornbill *Bucorvus leadbeateri* (Vigors) (Aves: Bucerotidae)." *Ann. Transvaal Mus.* 32:65–100.

Kennedy, E. D. 1991. "Determinate and Indeterminate Egg-laying Patterns: A Review." *Condor* 93:106–24.

Kenyon, K. W., and D. W. Rice, 1958. "Homing of Laysan Albatrosses." *Condor* 60:3–6.

Kilham, L. 1971. "Use of Blister Beetle in Bill-sweeping by White-breasted Nuthatch." *Auk* 88:175–76.

King, B. R. 1980. "Social Organization and Behaviour of the Gray-crowned Babbler *Pomatostomus temporalis.*" *Emu* 80:59–76.

Klem, D., Jr. 1990. "Collisions Between Birds and Windows: Mortality and Prevention." *J. Field Ornithol.* 61:120–28.

Lack, D. 1961. *Darwin's Finches.* New York: Harper and Brothers.

Levick, G. M. 1914. *Antarctic Benguins.* London: Robert McBride.

Lincoln, F. C. 1950. *Migration of Birds.* Circular 16, Fish and Wildlife Service, U.S. Dep. of the Interior. Washington, D.C.: U.S. Government Printing Office.

Lockley, R. M. 1942. *Shearwaters.* London: J. M. Dent.

Lorenz, K. Z. 1952. *King Solomon's Ring.* London: Methuen.

Lovell, H. B. 1958. "Baiting of Fish by a Green Heron." *Wilson Bull.* 70:280–81.

Manwell, R. D. 1964. "A Congenitally One-legged Cowbird." *Auk* 81:438–39.

Marks, J. S., and C. S. Hall. 1992. "Tool Use by Bristle-thighed Curlews Feeding on Albatross Eggs." *Condor* 94:1032–34.

Marshall, A. J. 1954. *Bower-birds: Their Displays and Breeding Cycles.* Oxford: Clarendon Press.

Matthews, G. V. T. 1968. *Bird Navigation,* 2nd. ed. Cambridge: Cambridge University Press.

Matthews, L. H. 1939. "Visual Stimulation and Ovulation in Pigeons." *Proc. Royal Soc. London,* ser. B, 126:557–60.

Mazzeo, R. 1953. "Homing of the Manx Shearwater." *Auk* 70:200–201.

Meyerriecks, A. C. 1972. "Tool-using by a Double-crested Cormorant." *Wilson Bull.* 84:482–83.

Millikan, G. C., and R. L. Bowman. 1967. "Observations on Galápagos Tool-using Finches in Captivity." *Living Bird* 6:23–41.

Moksnes, A., E. Røskaft, and A. T. Braa. 1991. "Rejection Behavior by Common Cuckoo Hosts toward Artificial Brood Parasite Eggs." *Auk* 108:348–54.

Montevecchi, W. A. 1978. "Corvids Using Objects to Displace Gulls from Nests." *Condor* 80:349.

Moore, F. R. 1990. "Evidence for Redetermination of Migratory Direction Following Wind Displacement. *Auk* 107:425–28.

Moreau, R. E. 1938. "A Contribution to the Biology of the Musophagidae, the So-called Plantain-eaters. *Ibis* for October:639–71.

Morse, D. H. 1968. "The Use of Tools by Brown-headed Nuthatches." *Wilson Bull.* 80:220–24.

Murray, B. G., Jr. 1966. "Migration of Age and Sex Classes of Passerines on the Atlantic Coast in Autumn." *Auk* 83:352–60.

Myres, M. T. 1964. "Dawn Ascent and Reorientation of Scandinavian Thrushes (*Turdus* spp.) Migrating at Night over the Northeastern Atlantic Ocean in Autumn." *Ibis* 106:7–51.

Nice, M. M. 1943. *Studies in the Life History of the Song Sparrow. II: The Behavior of the Song Sparrow and Other Passerines.* Trans. Linnaean Soc. N. Y. vol. 6.

Nolan, V., Jr. 1958. "Anticipatory Food-bringing in the Prairie Warbler." *Auk* 75:263–78.

———. 1978. "The ecology and behavior of the Prairie Warbler *Dendroica discolor.*" American Ornithol. Union, *Ornithol. Monogr.* No. 26.

Orenstein, R. I. 1972. "Tool-use by the New Caledonian Crow *(Corvus moneduloides).*" *Auk* 89:674–76.

Ortega, J. C., and M. Bekoff. 1987. "Avian Play: Comparative Evolutionary and Developmental Trends." *Auk* 104:338–41.

Pepperberg, I. M. 1985. "Social Modeling Theory: A Possible Framework for Understanding Avian Vocal learning." *Auk* 102:854–64.

———. 1990a. "Some Cognitive Capacities of an African Grey Parrot *(Psittacus erithacus).*" *Advances in the Study of Behavior* 19:357–409. London: Academic Press.

———. 1990b. "Cognition in an African Gray Parrot *(Psittacus erithacus):* Further Evidence for Comprehension of Categories and Labels." *Journ. Comparative Psychol.* 104:41–52.

———. 1990c. "Referential Mapping: A Technique for Attaching Functional Significance to the Innovative Utterances of an African Gray Parrot *(Psittacus erithacus).*" *Applied Psycholinguistics* 11:23–44.

Pepperberg, I. M., and M. V. Brezinsky. 1991. "Acquisition of a Relative Class Concept by an African Gray Parrot *(Psittacus erithacus):* Discriminations Based on Relative size." *Journ. Comparative Psychol.* 105:

Prescott, K. W. 1947. "Unusual Behavior of a Cowbird and Scarlet Tanager at a Red-eyed Vireo Nest." *Wilson Bull.* 59:210.

Preston, C. R., H. Moseley, and C. Moseley. 1986. "Green-backed Heron Baits Fish with Insects." *Wilson Bull.* 98:613–14.

Pruett-Jones, S. G., and M. A. Pruett-Jones. 1988. "The Use of Court Objects by Lawes' Parotia." *Condor* 90:531–45.

Ramsay, A. O. 1951. "Familial Recognition in Domestic Birds." *Auk* 68:1–16.

Rappole, J. H., and D. W. Warner. 1980. "Ecological Aspects of Migrant Bird Behavior in Veracruz, Mexico." In *Migrant Birds in the Neotropics: Ecology, Behavior, Distribution, and Conservation,* A. Keast and E. S. Morton, eds., 353–93. Washington, D.C.: Smithsonian Institution Press.

Richardson, W. J. 1978. "Reorientation of Nocturnal Landbird Migrants over the Atlantic Ocean near Nova Scotia in Autumn." *Auk* 95:717–32.

Ridley, G. N. 1952. *Your Brain and You.* London: Watts.

Ripley, D. 1940. *Trail of the Money Bird: 30,000 Miles of Adventure with a Naturalist.* New York: Harper.

Riska, D. E. 1984. "Experiments on Nestling Recognition by Brown Noddies *(Anous stolidus).*" *Auk* 101:605–9.

Ristau, C. A. 1991. Aspects of the Cognitive Ethology of an Injury-feigning Bird, the Piping Plover. In *Cognitive Ethology: The Minds of other Animals,* C. A. Ristau, ed. Hillsdale, New Jersey: Lawrence Erlbaum Associates, 91–126.

Roberts, B. B. 1934. "Notes on the Birds of Central and Southeast Iceland, with Special Reference to Food-habits." *Ibis* (13)4:239–64.

Rowley, I. 1983. Remating in Birds. In *Mate Choice,* P. Bateson, ed. Cambridge: University Press, 331–60.

———. 1990. *Behavioural Ecology of the Galah* Eolophus roseicapillus. Chipping Norton, N.S.W., Australia: Surrey Beatty.

Sabine, W. S. 1959. "The Winter Society of the Oregon Junco: Intolerance, Dominance, and Pecking Order." *Condor* 61:110–35.

Sargent, T. D. 1962. "A Study of Homing in the Bank Swallow *(Riparia riparia).*" *Auk* 79:234–46.

Schwartz, P. 1963. "Orientation Experiments with Northern Waterthrushes Wintering in Venezuela." *Proc. 13th Interntl. Ornithol. Congr.* 481–84.

———. 1964. "The Northern Waterthrush in Venezuela." *Living Bird* 3:169–84.

Selous, E. 1927. *Realities of Bird Life.* London: Constable.

Seymour, R. S., and D. F. Bradford. 1992. "Temperature Regulation in the Incubation Mounds of the Australian Brush-turkey." *Condor* 94: 134–50.

Sherman, A. R. 1952. *Birds of an Iowa Dooryard.* Boston: Christopher Publishing House.

Shy, M. M. 1982. "Interspecific Feeding among Birds: A Review." *J. Field Ornithol.* 53:370–93.

Sick, H. 1993. *Birds in Brazil: A Natural History.* W. Belton, trans. Princeton, New Jersey: Princeton University Press.

Simmons, R. 1992. "Brood Adoption and Deceit among African Marsh Harriers *Circus ranivorous.*" *Ibis* 134:32–34.

Skutch, A. F. 1953. "How the Male Bird Discovers the Nestlings." *Ibis* 95:1–37, 505–42.

———. 1954. *Life Histories of Central American birds,* vol. 1. Pacific Coast Avifauna 31. Berkeley, Calif.: Cooper Ornithological Society.

———. 1954–55. "The Parental Stratagems of Birds." *Ibis* 96:544–64; 97:118–42.

———. 1959. "Life History of the Groove-billed Ani." *Auk* 76:281–317.

———. 1960. *Life Histories of Central American birds,* vol. 2. Pacific Coast Avifauna 34. Berkeley, Calif.: Cooper Ornithological Society.

————. 1961. "Helpers among Birds." *Condor* 63:198–226.

————. 1969. *Life Histories of Central American birds,* vol. 3. Pacific Coast Avifauna 35. Berkeley, Calif.: Cooper Ornithological Society.

————. 1971. "Life History of the Keel-billed Toucan." *Auk* 88:381–96.

————. 1976. *Parent Birds and Their Young.* Austin: University of Texas Press.

————. 1980. *A Naturalist on a Tropical Farm.* Berkeley: University of California Press.

————. 1983a. *Birds of Tropical America.* Austin: University of Texas Press.

————. 1983b. *Nature through Tropical Windows.* Berkeley: University of California Press.

————. 1985. *Life Ascending.* Austin: University of Texas Press.

————. 1987a. *A Naturalist amid Tropical Splendor.* Iowa City: University of Iowa Press.

————. 1987b. *Helpers at Birds' Nests: A Worldwide Survey of Cooperative Breeding and Related Behavior.* Iowa City: University of Iowa Press.

————. 1989. *Birds Asleep.* Austin: University of Texas Press.

————. 1991. *Life of the Pigeon.* Ithaca, N.Y.: Cornell University Press.

————. 1992. *Origins of Nature's Beauty.* Austin: University of Texas Press.

Smith, G. A. 1971. "Further examples of 'tool-using' parrots." *Avicult. Mag.* 77:47–48. Abstract in *Ibis* 114:578 (1972).

Snow, D. W. 1958. "The Breeding of the Blackbird *Turdus merula* at Oxford." *Ibis* 100:1–30.

Sordahl, T. A. 1990. "The Risks of Avian Mobbing and Distraction Behavior: An Anecdotal Review." *Wilson Bull.* 102:349–52.

Southern, W. E. 1968. "Experiments on the Homing Ability of Purple Martins." *Living Bird* 7:71–84.

Sprunt, A., Jr. 1944. "Remarkable Aërial Behavior of the Purple Martin." *Auk* 61:296–97.

Stacey, P. B., and W. D. Koenig, eds. 1990. *Cooperative Breeding in Birds: Long-term Studies of Ecology and Behavior.* Cambridge: Cambridge University Press.

Stager, K. E. 1964. "The Role of Olfaction in Food Location by the Turkey Vulture *(Cathartes aura)."* *Contrib. in Sci.* no. 81:1–63. Los Angeles, Calif.: Los Angeles County Museum.

Stonehouse, B. 1953. "The Emperor Penguin *Aptenodytes forsteri* Gray. 1: Breeding behaviour and development." *Falkland Islands Dependencies Survey, Sci. Reports* 6:1–33. London: H. M. Stationery Office.

Stonehouse, B., and S. Stonehouse. 1963. "The Frigate Bird *Fregata aquila* of Ascension Island." *Ibis* 103b:409–22.

Stoner, E. A. 1947. "Anna Hummingbird at Play." *Condor* 49:36.

Taverner, P. A. 1936. "Injury Feigning by Birds." *Auk* 53:366.

Thomas, B. T. 1984. "Maguari Stork Nesting: Juvenile Growth and Behavior." *Auk* 101:812–23.

Thorpe, W. H. 1956. *Learning and Instinct in Animals.* London: Methuen.

Thouless, C. R., J. M. Fanshaw, and B. C. R. Bertram. 1989. "Egyptian Vultures *Neophron percnopterus* and Ostrich *Struthio camelus* Eggs: The Origins of Stone-throwing Behaviour." *Ibis* 131:9–15.

Thurber, W. A. 1981. "Aerial 'Play' of Black Vultures." *Wilson Bull.* 93:97.

Tinbergen, N. 1953. *The Herring Gull's World.* London: Collins.

Vehrencamp, S. L. 1977. "Relative Fecundity and Parental Effort in Communally Nesting Anis *(Crotophaga sulcirostris)*." *Science* 197:403–5.

———. 1978. "The Adaptive Significance of Communal Nesting in Groove-billed Anis *(Crotophaga sulcirostris)*." *Behav. Ecol. Sociobiol.* 5:1–33.

Vuren, D. V. 1984. "Aerobatic Rolls by Ravens on Santa Cruz Island, California." *Auk* 101:620–21.

Wall, S. B. V. 1990. *Food Hoarding in Animals.* Chicago, Ill.: University of Chicago Press.

Wallace, A. R. 1872. *The Malay Archipelago.* London: Macmillan.

Weidenreich, F. 1948. "The Human Brain in the Light of its Phylogenetic Development." *Scientific Monthly* 67:103–9.

Welty, J. C. 1975. *The Life of Birds,* 2nd. ed. Philadelphia: W. B. Saunders.

Whitford, P. C. 1990. "Deception in Canada Geese." *Wilson Bull.* 102:558–59.

Wilkinson, R. 1982. "Social Organization and Communal Breeding in the Chestnut-bellied Starling *(Spreo pulcher)*." *Anim. Behav.* 30:1118–28.

Willis, E. O. 1972. "The Behavior of the Spotted Antbird." Amer. Ornithol. Union, *Ornithol. Monogr.* no. 10:i–vi, 1–162.

Zahavi, A. 1990. "Arabian Babblers: The Quest for Social Status in a Cooperative Breeder." In *Cooperative Breeding in Birds,* P. B. Stacey and W. D. Koenig, eds., 105–30.

Index

Note: Pages with illustrations are indicated by italics.